This volume demonstrates the bewildering social
ciated with the adoption of internet technology in Latin America.
This volume convincingly shows that technology is always intricately
bound up with the social structures in and through which it operates. A
unique engagement between technology studies and area studies. Latin
America presents an unusual combination of obsolete and cutting-edge
technology environments, which conspire to generate the fascinating
paradoxes described in this outstanding volume.

Dr. Jochen Kleinschmidt, Universidad del Rosario, Colombia.

Plaw, Carvalho, and Ramírez Plascencia have assembled an exciting
and diverse group of scholars to address important questions including:
What are the effects of the increased use of the Internet by people in
Latin America? What are the challenges faced by privacy, data mining,
and cyberbullying? What can be the best national responses and policies? How should we understand increased automation in industrial
production? What is the effect of the use of drones in military campaigns, policing, drug trafficking, healthcare provision, and leisure?
What are the effects of online information bubbles on social media on
democratic outcomes? How is social media changing social movements
and contentious politics? This volume forces us to think creatively and
ethically about the challenges faced by technological change. It brings
voices from Latin Americans based there as well as Latin Americans
based in the United States and Europe. Though, the relevance of the
discussions and implications go beyond Latin America.

Ernesto Castañeda, PhD, Associate Professor,
Department of Sociology, American University,
DC; Co-author of 'Social Movements 1768–2018.'

The Politics of Technology in Latin America (Volume 1)

This book analyzes the arrival of emerging and traditional information and technology for public and economic use in Latin America. It focuses on the governmental, economic, and security issues and also on the study of the complex relationship between citizens and government.

The book is divided into three parts:

- 'Digital data and privacy, prospects, and barriers' centers on the debates among the right to privacy and the loss of intimacy on the Internet,
- 'Homeland security and human rights' focuses on how novel technologies, such as drones and autonomous weapons systems, reconfigure the strategies of police authorities and organized crime,
- 'Labor Markets, digital media, and emerging technologies' emphasize the legal, economic, and social perils and challenges caused by the increased presence of social media, blockchain-based applications, artificial intelligence, and automation technologies in the Latin American economy.

This first volume in a two-volume set will be important reading for scholars and students of governance in Latin America, the protection of human rights and the use of technology to combat crime, and the new advances of digital economy in the region.

Professor Avery Plaw specializes in political theory and international relations, with a particular focus on strategic studies.

Barbara Carvalho Gurgel has a bachelor's degree in political science from the University of Massachusetts Dartmouth (USA), and is working toward a master's degree in journalism from the Harvard Extension School (USA).

David Ramírez Plascencia is a professor at the University of Guadalajara, specializing in the study of information law and digital policies.

Emerging Technologies, Ethics, and International Affairs
Series Editors: Steven Barela, Jai C. Galliott, Avery Plaw,
Katina Michael

This series examines the crucial ethical, legal, and public policy questions arising from or exacerbated by the design, development, and eventual adoption of new technologies across all related fields, from education and engineering to medicine and military affairs. The books revolve around two key themes:

- Moral issues in research, engineering, and design
- Ethical, legal, and political/policy issues in the use and regulation of Technology

This series encourages submission of cutting-edge research monographs and edited collections with a particular focus on forward-looking ideas concerning innovative or as yet undeveloped technologies. Whilst there is an expectation that authors will be well grounded in philosophy, law, or political science, consideration will be given to future-orientated works that cross these disciplinary boundaries. The interdisciplinary nature of the series editorial team offers the best possible examination of works that address the 'ethical, legal and social' implications of emerging technologies.

For more information about this series, please visit: https://www.routledge.com/Emerging-Technologies-Ethics-and-International-Affairs/book-series/ASHSER-1408

Social Robots
Boundaries, Potential, Challenges
Marco Nørskov

Legitimacy and Drones
Investigating the Legality, Morality, and Efficacy of UCAVs
Steven J. Barela

American Security and the Global War on Terror
Edwin Daniel Jacob

The Politics of Technology in Latin America (Volume 1)
Data Protection, Homeland Security, and the Labor Market
Edited by Avery Plaw, Barbara Carvalho Gurgel, and David Ramírez Plascencia

The Politics of Technology in Latin America (Volume 2)
Digital Media, Daily Life, and Public Engagement
Edited by David Ramírez Plascencia, Barbara Carvalho Gurgel, and Avery Plaw

The Politics of Technology in Latin America (Volume 1)

Data Protection, Homeland Security and the Labor Market

Edited by Avery Plaw, Barbara Carvalho Gurgel, and David Ramírez Plascencia

Routledge
Taylor & Francis Group

LONDON AND NEW YORK

First published 2021
by Routledge
2 Park Square, Milton Park, Abingdon, Oxon OX14 4RN

and by Routledge
52 Vanderbilt Avenue, New York, NY 10017

Routledge is an imprint of the Taylor & Francis Group, an informa business

British Library Cataloguing-in-Publication Data
A catalogue record for this book is available from the British Library

Library of Congress Cataloging-in-Publication Data
A catalog record has been requested for this book

ISBN: 9780367359416 (hbk)
ISBN: 9780429342745 (ebk)

Typeset in Times NR MT Pro

by KnowledgeWorks Global Ltd.

This editorial project is dedicated to the memory
of Dr. Avery Plaw.

Contents

List of figures

List of tables

Contributors

Editors

Professor Avery Plaw specializes in political theory and international relations, with a particular focus on strategic studies. He is the author of *Targeting Terrorists: A License to Kill?* (Routledge, 2008) and the editor of *The Metamorphosis of War* (Brill, 2012) and *Frontiers of Diversity: Explorations in Contemporary Pluralism* (Brill, 2005). He is also the co-author of the textbook *The Drone Debate: A Primer on the U.S. Use of Unmanned Aircraft Outside Conventional Battlefields* (Rowman & Littlefield, 2015).

Barbara Carvalho Gurgel has a bachelor's degree in political science from the University of Massachusetts Dartmouth (USA), and is working toward a master's degree in journalism from the Harvard Extension School (USA). Her latest pieces involve such subjects as violence against journalists, online radicalization, and literary analysis of Childish Gambino's Grammy-winning record *This Is America*. Barbara has worked on various research projects and publications regarding extra-territorial uses of force, and the history and practice of drone strikes by the United States. Her published work *Super Soldiers: The Ethical, Legal and Social Implications* with Avery Plaw was part of Ashgate Publishing's *The Super Soldier as Scholar: Cultural Knowledge as Power.* She is currently working as lead editor at the Israeli Center for the Study of Targeted Killing (ICSTK), a project mapping the targeted killing throughout the Middle East. Barbara has also published her prose and poetry.

David Ramírez Plascencia is a Professor at the University of Guadalajara, specializing in the study of information law and digital policies. His previous works include "Concerning at distance: digital activism and social media empowerment between Latin-American migrants in Spain" in the journal *ESSACHESS* and "Education reform, teacher resistance and social media activism in Mexico, 2013–2016" in the journal *The Educational Forum*, Taylor & Francis. His research has focused on the study of the possibilities and dilemmas of information technologies in developing countries. He is interested in digital activism, information control, and social media use in Latin America.

Contributors

Amaranta Alfaro is researcher at the Journalism Department at Universidad Alberto Hurtado. She is a PhD candidate at the Graduate School of Media & Communication at Hamburg University, and is a sponsored student at the Centre for Social Conflict and Cohesion Studies (COES). Her doctoral research focuses on civic engagement through social media in Chile and seeks to explore its potential to reinforce social cohesion. Her other research interest are digital activism, digital inequalities, and digital journalism. She holds a master's degree in media, communication, and cultural studies from the universities of Roskilde (Denmark) and Kassel (Germany).

Matías Dodel is an Associate Professor of communication at Universidad Católica del Uruguay. He is the director of the Internet of People (IoP) research group, where he coordinates the Uruguayan chapters of international comparative Internet studies such as World Internet Project, From Digital Skills to Tangible Outcomes (DiSTO), and Global Kids Online. He holds a PhD from the Department of Sociology, University of Haifa. His research interests are digital inequalities, social stratification, digital safety, and cybercrime.

Patricio Cabello is an Assistant Professor at the Institute for Advanced Research in Education at Universidad de Chile and the Head of Kids Online Chile. He holds a PhD in social psychology (UCM, Spain), a master's in anthropology and development (U. de Chile), a master's in social research methods (UCM, Spain), and a master's in psychology (UCEN, Chile). His research topics are the children's right to communication, innovation in education, digital inclusion, and youth participation though digital media

Hugo Claros is a quantitative sociologist, specializing in information science. He is currently working toward a Master of Science degree in applied statistics. He has worked in the Ministry of Education of Peru in knowledge management, monitoring and evaluation, and information systems.

Edgar A. Ruvalcaba-Gomez is a Research Professor at the Universidad de Guadalajara (UDG), at the Center for Economic and Administrative Sciences, which is affiliated to the Research Institute in Public Policies and Government. He holds a PhD in law, government and public policy from Autonomous University of Madrid (UAM), Spain, a master's degree in public policy, and another master's degree in public management in virtual environments, both from UDG, México. He also holds a bachelor's degree in law from UDG. Edgar conducts research related to open government, transparency, citizen participation, open data, corruption, digital government and artificial intelligence, and public innovation. He has been a visiting scholar at Trinity College in Dublin, Ireland, and at the Center for Technology in Government (CTG), University at Albany, New York, USA. He has been a researcher for the Open Government Partnership in the Subnational Program and coordinator of the Open Government Academic Network in Mexico.

Soledad Gattoni is a researcher at the Center for Sociopolitical Studies at IDAES, UNSAM. She holds a PhD in social sciences from the University of Buenos Aires (UBA), a master's degree in political science from the University of Salamanca (USAL), and a bachelor's degree in political science from UBA. With more than 15 years of experience in the areas of transparency, citizen engagement, and accountability, she currently works as a researcher and consultant at the Independent Reporting Mechanism (IRM) of the Open Government Partnership, and at the World Bank's Governance Global Practice Department. Among other partners, she has worked as a consultant for Open Society Foundations, CIVICUS, Transparency International, and the Argentinian National Planning Ministry. She has been a Fulbright fellow and received scholarships from the Deutscher Akademischer Austaushdienst (DAAD), the Carolina Foundation, the CONICET, and UBA.

Raymond W. Weyandt is a Master of Public Affairs candidate at the LBJ School of Public Affairs at the University of Texas at Austin and an experienced public policy researcher, consultant, and photographer. His recent work focuses on open government, police accountability, migration, social equity, and human rights. Weyandt currently serves as research director for the Peace Mill and communications director for Innovations for Peace and Development, an international development research lab at UT Austin. He worked previously as a research consultant for the Open Government Partnership's Independent Reporting Mechanism.

Raúl Salgado Espinoza has a PhD in political science and international studies from the University of Birmingham, UK and a master's in political science from the University of Bonn, Germany. He is a full-term professor of International Studies at FLACSO, Ecuador and his main areas of teaching and research are foreign policy analysis, political thought and ethics, and international relations and small states and regionalism. His most recent publications are "Constructing Realities in International Politics: Latin American Views on the Construction and Implementation of the International Norm Responsibility to Protect (R2P)." In Steel, B. and Heinze, E. (eds.). 2018. *Routledge Handbook of*

Ethics and International Relations. Routledge; and *Small builds Big. How Ecuador and Uruguay* contributed to the construction of the UNASUR, 2017. Quito: Flacso, Ecuador.

David S. Dalton is an Assistant Professor of Spanish at the University of North Carolina, Charlotte. His research focuses primarily on how science, medicine, and technology interface with constructs of race, class, and gender in Mexico. His previous books include *Mestizo Modernity: Race, Technology, and the Body in Postrevolutionary Mexico* (U of Florida P, 2018) and "Ibero-American Zombies: Undead Colonialism," a special edition to *Alambique* (vol. 6, no. 2, 2018), which he coedited with Sara Potter. He has published 15 articles in major journals of Hispanic and Latin American studies. He is currently coediting two volumes that deal with scientific thought in Mexico and throughout Latin America. One is a reader on healthcare in Spanish America, while the other is a book on irreligion and the secular tradition in Mexico. He is also working on his next monograph, *Robo Sacer: Mexico, Technology, and Border-Crossing Resistance*, which interrogates how technology validates the United States-Mexico border as an exceptional state that dehumanizes Mexicans on the one hand while providing opportunities for resistance on the other.

Maximiliano Marzetti is Assistant Professor of law at IÉSEG School of Management and member of the IÉSEG's Center of Excellence in Negotiation in Paris. In addition, he is adjunct faculty at ESCP EUROPE – Paris, the University Institute of European Studies at Turin University, the Latin American School of Social Sciences (FLACSO) – Argentina, and coordinates Harvard University's CopyrightX (a Turin University affiliate course) and World Intellectual Property Organization's distance learning courses. Maximiliano was awarded an Erasmus Mundus grant by the European Commission, a semi-senior research grant by the Latin American Council of Social Sciences (CLACSO), was three times visiting scholar at the Max Planck Institute for Innovation and Competition in Munich, was appointed Associate Researcher at the Centre of Interdisciplinary Studies of Industrial and Economic Law (CEIDIE) at the University of Buenos Aires, and was, formerly, the Deputy Director to the LLM in Intellectual Property at Turin University, in collaboration with the World Intellectual Property Organization. He had served as a trial lawyer, commercial mediator, and industrial property agent in Buenos Aires. Maximiliano obtained a law degree from the Pontificia Universidad Católica of Argentina, earned two master of law degrees, one from Turin University in Intellectual Property, another from Hamburg and Bologna Universities in Law and Economics (dual degree), and a PhD from Erasmus Rotterdam Universiteit.

Victoria Basualdo has a PhD in history from the University of Columbia. She is a researcher at CONICET, coordinator of the Program "Labor studies, trade union movement and industrial organization" in the area of Economics and Technology of FLACSO, Argentina, member of IICSAL-CONICET and Professor of the Master in Political Economy (FLACSO) and in other academic institutions. She is the author of numerous articles and chapters in academic publications, co-author of the book *La Industria y sindicalismo de base en la Argentina* (Cara o Ceca, 2010), and coordinator of the books *Recent Transformations in the Argentine Economy* (Prometeo, 2008) and *La Argentine Working Class in the 20th Century: Forms of Struggle and Organization* (Cara o Ceca, 2011), *Labor Outsourcing: Origins, Impact and Keys for Its Analysis in Latin America* (XXI century, 2014), *Outsourcing and Labor Rights in Argentina Current* (National University of Quilmes, 2015), and was one of the coordinators of the book *Corporate Responsibility in Crimes Against Humanity: Repression of Workers During State Terrorism* (Infojus, 2015), prepared by a team of FLACSO, CELS, the Truth and Justice Program and the Secretariat of Human Rights.

Graciela Bensusán is Research Professor at the Universidad Autónoma Metropolitana-Xochimilco in Mexico City, a post she has held since 1976. Since 1989, she has also been Professor (part-time) at the Facultad Latinoamericana de Ciencias Sociales-México. She holds a law degree from the Facultad de Derecho y Ciencias Sociales at the Universidad de Buenos Aires and a PhD in political science from the Facultad de Ciencias Políticas y Sociales at the Universidad Nacional Autónoma de México. Professor Bensusán is a member of Mexico's Sistema Nacional de Investigadores (Level III). She has also held diverse research appointments at the Economic Commission for Latina America and Caribe in Santiago, Chile; the Institute of the Americas – University College of London; and King's College in UK. Professor Bensusán is author of *El modelo mexicano de regulación laboral* (2000, Miguel Angel Porrúa- UAM) and co-author of *Sindicatos y Política en México: cambios, continuidades y contradicciones* (2013, UAM, FLACSO, and CLACSO), *Diseño legal y desempeño real: instituciones laborales en América Latina* (2006, UAM- Miguel Angel Porrúa), and editor or coeditor of twenty other books. Her coedited book *Trabajo y trabajadores en el México contemporáneo* (2000) won the Latin American Studies Association Labor Section's book prize in 2001. Professor Bensusán's current research focuses on the comparative analysis of labor policies, institutions, and organizations in Latin America.

Dasten Julián-Vejar has a PhD in sociology from the Friedrich Schiller Universität de Jena (Germany). He is an academic and researcher in the Department of Sociology and Political Science, and Director of the Work Studies Group from the South (GETSUR) at the Universidad Católica de Temuco (Chile). Julián-Vejar is also Associate Researcher at the Society, Work & Politics Institute, University of Wittwatersrand (South Africa).

He is a researcher at the National Commission of Science and Technology (CONICYT), Chile and at Deutscher Akademischer Austauschdienst (DAAD). He is a specialist in labor relations, precarious work, and social movements and is member of the Latin American Association of Sociology of Work.

Carmina Rodríguez-Hidalgo is an Assistant Professor at the School of Communications and Journalism at Universidad Adolfo Ibáñez. She holds a PhD in communication from the University of Amsterdam and is a Research Master of Science in Communication, here. She is also a journalist from the University of Chile. Her recent work focuses on the theoretical, socio-emotional outcomes, and current applications of social robotics. In general, her expertise is within the psychological outcomes of technology use and mediated communication, particularly in children and youth. Her doctoral thesis focused on the emotional effects of sharing emotions online. Twitter: @Carmi_Rodrig.

Rosa María Alonzo González is Professor at University of Guadalajara, and collaborates with the research group Agorante, Universidad de Colima. Rosa María holds a doctoral degree in social sciences and is a candidate for the membership of the National Research System of Mexico (Sistema Nacional de Investigadores, SNI) She does research in the social uses of information and communication technologies (ICT). Her previous works include "Ending the Digital Gender Divide. Are Coding Clubs the Solution?" (Trípodos, 2020) and "Sobre el uso de los conceptos prosumer y emirec en estudios sobre comunicación" (Sphera, 2019). She is interested in studying about sharing economy, social media consumption, and online discourses.

1 Introduction

*Avery Plaw, Barbara Carvalho Gurgel,
and David Ramírez Plascencia*

The popularization of the Internet, beginning in the last decade of the 20th century, and accelerated by the growth of mobile technology in the first decade of the 21st, has had a huge impact around the world. This change not only affects how people communicate, shop, and entertain, but also the way they see the world and how they represent themselves. In addition, the huge advances in the field of artificial intelligence and robotics have led to the invention of sophisticated social robots that are capable of mimicking humans and creating social relationships with people. The impact of these developments has been widely noted and discussed in the most technologically advanced countries of the world, which are often at the cutting edge of emerging trends, but they have been far less carefully examined in regard to the developing world, and Latin America in particular.

This oversight is unfortunate, because emerging technologies are having an enormous impact, and raising equally difficult (and often distinctive) issues in Latin America. The use of the Internet in general, and social media in particular, is one of the most significant social activities for inhabitants of the region, despite uneven accessibility (Statista Research Department, 2020b). New information technologies are continuously arriving in the region as well, from the Internet of things to artificial intelligence, drones, robots, all of which are conspiring not only to change the domestic, social, and economic environment, but to transform social and economic links across the region and externally with the outside world. In a real sense, the Internet has cut the region's geographical moorings and plunged it and its inhabitants into an integrated and multi-tiered global virtual space.

Moreover, this technological omnipresence has raised novel legal, ethical, and political issues concerning the relationship between the governments and their citizens, especially regarding the contraposition of negative and positive liberties. The protection of civil rights such as free speech and privacy comes up against the governments' efforts to monitor user activity online (as a way to protect citizens from, for example, terrorists and criminal gangs that operate transnationally). At the international level, many pressing issues have arisen such as foreign government intervention in regional elections through the use of bots, fake news, hacking, and other informational war strategies. Locally, some matters like the use of unmanned drones to combat crime and insurgent movements across Latin America are raising public concerns as well. Actually, there is a growing temptation to impose deep technical and legal controls on the immense flows of virtual data on the Internet. However, history reveals that trying to find a balance amongst security and freedom is not an easy task. What is clear around the world is that despite deep disagreements about whether and how to regulate new technologies, things have changed in ways that demand careful public attention and potential reform. The Latin American context also raises some distinctive issues in regard to the diffusion of these new technologies. The economic agenda, for example, is marked by a labor transition from the industrial model to the knowledge economy, in which both rural and urban production sectors will be impacted by the use of robots, automation, and connectivity, raising concerns about how these emerging technologies will reshape the labor market in countries like Mexico, Brazil, and Argentina.

In 2020, these concerns reached a next level since the outbreak of the COVID-19 global pandemic. Authorities, even in Latin America, have made use of digital devices such as drones, robots, and mobile applications to support their fight against the disease. Drones and robots have committed to undertaking surveillance tasks, sanitizing public areas, and advising people to avoid crowds and to stay at home (EFE, 2020). Other technologies such as phone location tracking have been employed to enforce quarantine and monitor the spread of the virus. However, besides the potential benefits of using these devices to combat the pandemic, the incorporation of these technologies compromise users' sensitive data such as identity, preferences, associations, and opinions (Human Rights Watch, 2020). This fact is particular risky in countries without robust democratic institutions and regulations, exposing users to misuse of their private data, and even political persecution. While understanding the real impact of employing these applications to diminish the effects of the pandemic is yet under development, there is no doubt that digital media and smart phones are no longer an accessory; those technologies play a key role in people's lives that cannot simply be overlooked, particularly regarding the protection of human rights.

The structure of the book

The politics of technology in Latin America (Volume 1): data protection, homeland security and the labor market

This book focuses on the analysis of the complex relationship between citizens and the government in an interconnected context. Volume 1 is divided into three main sections. Part I. Digital data and privacy, prospects and barriers. Part II. "Homeland security and human rights, a questioned balance?" and Part III. "Labor Markets, digital media and emerging technologies:

potentials and risks." Every section embraces one key topic related with the control and regulation of digital technologies in Latin America. The first part, "Digital data and privacy, prospects and barriers" focuses on the academic and public debates about of the right of privacy and the loss of intimacy in the Internet, where the user's private data is exposed, not only to criminal organizations, but to commercial and government use as well. Part II. "Homeland security and human rights, a problematic balance?" focuses on how novel technologies such as drones and autonomous weapons systems reconfigure the strategies of police authorities and organized crime. In Part III. "Labor Markets, digital media and emerging technologies: potentials and risks," the discussions emphasize the legal, economic and social perils and challenges caused by the increased presence of social media, blockchain-based applications, artificial intelligence and automation technologies in the Latin American economy. What would be the future landscape of labor markets in Latin America with the arrival of theses novel technologies?

Part I. Digital data and privacy, prospects, and barriers

As the adoption of information technologies increases around the world, there are growing uncertainties about how the government and the corporations gather and manipulate user private data for commercial, electoral, and policymaking purposes, and what kinds of legal and ethical frameworks could be proposed to protect users' private data? Actually, customer data has become one of the most important economic assets. Corporations use this information to develop new services and products, and to create efficient forms to merchandize their goods. A great portion of Facebook and Google's revenues rely on the commercialization of information shared by their users inside their platforms (Taylor, 2020). Public agencies use private data to design policies and to propose novel regulations and procedures, and political parties and candidates contract marketing agencies to know electoral preferences by analyzing social media. However, there are potential hazards in how governments and corporations could misuse private data for surveillance activities or to indiscreetly sell sensitive data for advertising, as it happened in the well-known case of Facebook and Cambridge Analytica. Besides the importance of digital technologies for the global economy, and the growing market for digital assets and services in Latin America, the regulation of ecommerce, intellectual property, or private data in many countries is outdated and contradictory. In addition, the procedure to comply with the law and to pursue criminals is practically inoperant (Heinemeyer, 2019). At a user level, there is an upward apprehension towards how mobile applications and platforms could compromise people's privacy, and how people are unprotected from electronic scams: the stealing of information to create fake accounts, spamming bots, and malicious actors who approach the users to obtain private data and commit fraud or extortion (Blankstein & Romero, 2019).

This topic is particularly relevant considering the privacy of children and young people who are more susceptible to being affected, when using digital media. There were even circumstances where little children are vulnerable, even before they were capable of using an electronic tablet. This is the case of "smart toys," sophisticated dolls or plush bears which are able to record, answer, and process information between the children and the companies which develop these products (Yadron, 2016). These kinds of cases have triggered the public debate on the potential risks of using these technologies among children: from cyberbullying to grooming or sexting, or even exposure to inappropriate images, the loss of privacy and the misuse of their data (Correa, 2016). These discussions could be appreciated in more detail in Alfaro, Dodel and Cabello's chapter "The reception of sexual messages among young Chileans and Uruguayans: Predictive factors and perception of harm," which focuses on how children and teenagers are exposed to the exchange of sexual content through digital media. The other two chapters in this section debate the proper management of private and public data. They explore how governments are contending with the management of data and the settlement of policies to protect the integrity of people's information and to increase governmental transparency. Hugo Claros in "Small Data, Big Data and the Ethical Challenges for a fragmented developing world: Peru's need for diversity-aware public policies on information technologies and practices," focuses on the analysis of the Peruvian case to explain how it is mandatory to democratize the promises to acquire value from data but, at the same time, to set a proper procedure that protects citizens from governmental and corporational misuse. A potential framework to guarantee the fair access to public information could stand, as Ruvalcaba, Gattoni, and Weyandt assert in their chapter "Open Government, Dilemmas, and Innovation at the Local Level: Comparing the Cases of Austin, Buenos Aires and Madrid, " on the implementation of models that encourage innovation inside the public administration allowing access to public data, fighting corruption, and empowering citizens. But as it will be possible to observe in this section, despite efforts to implement efficient channels to manage and share public data and to improve transparency and accountability, there are still concerns about the lack of regulations, controls, and proper management systems that protects citizens' data from potential misuse and exploitation.

Part II. Homeland security and human rights, a questioned balance?

Modern military forces have come to employ a mixed formula which includes diverse traditional and novel elements including robust intelligence services (employing, for example, vast systems of surveillance and electronic interception) to most effectively utilize more traditional ground forces (infantry and armaments) and coordinate close air support. In addition, governments and dissenting organizations organize information warfare campaigns to hack public agencies, steal information, and blackmail companies and governments (Rempfer, 2019). Organized cyber-attacks against foreign countries are a very common strategy to get classified information, damage computer systems that control oil refineries and airports, and even to directly influence the outcome of foreign presidential elections thanks to the spread of fake news supporting or defaming candidates (Vosoughi et al., 2018). The incorporation of information technologies into the development of new tactics and armaments has triggered a race among the global superpowers to produce novel weapons and systems that provide significant advantages towards contemporary and prospect adversaries.

This race to innovate and bring "disruptive technologies" into the battlefield is well-illustrated by the case of Unmanned Aerial Vehicles (UAVs), commonly known as "drones." Drones have been very successful in the security and military sectors. These devices are able to support sophisticated civil and public activities like surveillance, package distribution, and information gathering. In the frontline they are able to carry missiles and destroy targets at a distance (Dunn, 2013). In recent years, the use of armed predator, reaper, and other drones to target military and terrorist objectives has increased (Plaw et al., 2015) mainly because some countries have found them to be a cost-effective option both when killing terrorist leaders and when undertaking high risk missions that might otherwise endanger military personnel and civilians. There have also been recent cases in which cheap modified commercial drones have been used to threaten the lives of Latin American public figures – most famously Venezuelan Nicholas Maduro on August 4, 2018 (Levin & Beene, 2018). They also have been used by non-state actors to kill soldiers and to damage expensive military infrastructure (The Economist, 2018).

The more frequent use of drones and other emerging technologies by police forces in Latin America creates important legal challenges particularly regarding the protection of human rights and the sets of normative mores that secure the proper usage of drones to combat organized crime organizations that have access to strong firepower that are oftentimes equal to those used by the army (Cawley, 2017). In fact, the use of drones and of lethal autonomous weapon systems (LAWS) for political, security, and war-waging purposes have triggered intense concerns ranging from individual to national security, and even to regional and global stability. Raúl Salgado analyzes in his chapter "Ethical controversies about Lethal Autonomous Weapons Systems: views of small South American States," the ethical and legal dilemmas of incorporating LAWS for surveillance and defense activities, particularly the potential risk of causing the deaths of civilians. The author provides a strong analysis about the debates regarding these technologies in two Latin American countries, Ecuador and Uruguay. David Daltons' chapter, "From Sensationalist Media to the Narcocorrido: Drones, Sovereignty, and Exception along the US-Mexican Border," on the contrary, emphasizes how the incorporation of these devices has not only affected the relationship between the public forces and the narco-gangs in the US-Mexican border, but they have consolidated as objects of the cultural imagination and the narrative about narcotrafficking and migration across both sides of the border.

While technologies like drones or LAWS hold the potential of improving homeland security, they also invite misuse by criminal organizations. In Latin America, criminal organizations have been enthusiastic to take advantage of these technologies. They have adopted technologies to consolidate and spread their operations across the globe: to recover the gains through sophisticated money-laundering transnational systems, to improve the logistics of their shipments and the distribution of the product to the final consumers, and to monitor and receive tactical information about the activities of the authorities and other cartels. They have even adopted drones and autonomous vehicles, such as submarines, to carry thousands of pounds of drugs across national borders. This process of "technologization" inside criminal gangs presents an important challenge for local and international security agencies. Authorities must face the problem of fighting narco-cartels and criminal transnational organizations that have found shelter inside dark virtual networks (Angelini & Gibson, 2007). Inside these spaces, criminals and terrorists operate with a certain level of freedom. This lack of control is caused mainly by two important facts: (a) the fast development and sophistication of digital technologies such as encryption that enables the existence of these unregulated spaces looking to avoid police surveillance (Keene, 2011), and (b) the scarcity of a proper legal framework in cyberspace, not only to allow authorities to combat cybercrime but to bring criminals before the court. The last chapter of this section, "The process of technologization of the drug war in Mexico" by Avery Plaw, David Ramírez, and Barbara Carvalho provide a thorough analysis of the process of "technologization" of the drug war in Mexico. It describes what kinds of technologies are used, but also how this process has shaped the actions and strategies of the diverse actors involved: citizens, civil organizations, governments, and the narco-cartels.

Part III. Labor markets, digital media, and emerging technologies: potentials and risks

Robotics and artificial intelligence are no longer subjects of novels and sci-fi series but vibrant research fields that are expanding their presence in labor, academic, and military sectors. Universities, governments, and private funds are making increasing investments into creating prototypes that could work with fabrics or in outer space: assisting senior citizens, serving as pets, helping doctors to execute complex surgeries, and of course, on the battlefield. Advanced robotics and similar emerging technologies are already producing huge social and urban changes in large metropoles across the globe. Ultimately, artificial intelligence systems will manage public services, providing enhanced solutions in real-time to complex metropolitan challenges like traffic control, infrastructure preservation, water supply, electricity, air pollution, and security. Artificial Intelligence has been incorporated into the development of treatments for terrible diseases like cancer or AIDS, the control of domestic appliances at home, and supplying basic information in the case of AI-powered devices like the Amazon Echo. In Latin America there has been a growing interest in both academic research on, and commercial applications of, robotics as well as on popular educational activities, such as national and regional competitions among students in robotics, computer coding and applied mathematics (Ruiz-del-Solar & Weitzenfeld, 2012). By now, it is not uncommon to find Latin American students in robotic combat competitions or programmers working in companies developing software. More about this conversation on the potential uses and legal concerns of implementing AI technologies in Latin America could be consulted in Maximiliano Marzetti's chapter "Algorithmic Regulation – A legal framework for Artificial Intelligence in Latin America."

But besides these advances, there are growing reservations about how the expansion of robotics in Latin American industry, especially in the manufacturing sector, could cause massive loss of jobs. Governments are thus faced with the difficult challenge of trying to manage the promise of these new technologies while mitigating their all too tangible harms in the context of participating in vast global markets whose dynamics are almost entirely beyond their control. Moreover, these challenges are particularly exacerbated in Latin American societies which are already marked by relatively high levels of dependence on low-skilled labor, unemployment, poverty, and reliance on export to global markets. For many people the substantial automatization of industry and labor compromises employment and labor conditions. A massive influx of labor robots will produce the eventual loss of thousands of jobs, particularly in low-skill areas. Many developing countries that are dependent on manufacturing will suffer the worst impact of this transformation (Statista Research Department, 2020a). Robots and artificial intelligence systems are not just displacing humans in manufacturing and other repetitive labors (Berg et al., 2016), but thanks to the development of intricate systems that mix AI with robotics, machines are able to work as drivers, caregivers, virtual teachers, customer service assistants, and so on. These concerns about the automatization of production in Latin America is analyzed by Basualdo, Bensusan, and Julian-Vejar in their chapter "Automation and robotization of production in Latin America: problems and challenges for trade unions in the cases of Argentina, Mexico and Chile," the authors provide a comparative study of three countries in Latin America. This chapter contains crucial questionings such as, how to harmonize technological advances with corporate and worker needs? And how to face the potentially negative effects of the introduction of automatization and robotization in the Latin American manufacturing sector?

But the discussion about robots does not end with the economy. In the present day, robots are able to share everyday activities with humans, including senior citizens, people with disabilities and even children with autism, and currently they have even been incorporated to fight the COVID-19 pandemic. During 2020, many cities around the world used robots to sanitize surfaces and surveil the observance of public health recommendations such as forced quarantines and the congregation of crowds. As this editorial project is in development, Latin America, as every other region in the world, is suffering the effects of the COVID-19 pandemic. The expansion of the disease is bringing (and it will bring) important changes in the future for social and economic relations in the region. This context has encouraged the introduction of new technologies like drones and robots to support public health and the containment of the disease. In the chapter "Using functional and social robots to help during the Covid19 pandemic: Looking into the incipient case of Chile and its future artificial intelligence policy," Carmina Rodríguez-Hidalgo analyzes how Chile is considering the incorporation of robots to combat the spread of the pandemic. They could be important allies preventing infections, sanitizing nursing homes, and monitoring the observance of quarantines.

Along with the rising presence of robots and artificial intelligence, there are other markets related to digital media that are expanding in Latin America as well. This is the case of streaming services, as in the case of HBO, Netflix, or YouTube which are now in growing to the detriment of traditional services such as television (Ceurvels, 2019). This current trend is more noticeable regarding children and youth, which are enthusiastic consumers of multimedia content shared on the Internet or on mobile applications. According to a recent study (Statista Research Department, 2020a), more than 96% of responding parents of children aged three to 13 years old from Chile stated that their kids had access to devices with an internet connection, in Mexico the percentage is 83%, less than the South American country, but very impressive considering that the percentage of users in 2018 was only 65%. For many children most of their socialization flourishes in virtual spaces. They stay in contact with their friends and relatives using smartphones and applications such as Snapchat, WhatsApp, YouTube, or Instagram. Teenagers are great consumers of user-generated content distributed by Youtubers and social media influencers (Bailey et al., 2018). This entertainment sector has risen as an important source of revenue, many times the Youtuber is sponsored by big commercial corporations as the case of the Chilean Youtuber *Yo soy German* (I am German), or the Mexican influencers called *Yuya* and *The Polinesios* whose annual revenues could reach millions of US dollars. More about the economic and legal effects of these new markets of media consumption in Latin America can be consulted in Alonzo González's chapter "Intellectual property and social media policies for user generated content: some lessons from Mexico."

In the last chapter of this section, "Mining as an Art of Survival in Venezuela: Eluding Scarcity and improving Social Conditions with Bitcoins," David Ramírez analyzes the arrival of cryptocurrencies, mainly Bitcoin, to Latin America and how this virtual currency has become an important element in the economy in Latin America, particularly in countries such as Argentina and Venezuela in where people use Bitcoins as a method to save their money from the frequent devaluation of the national currencies, and it is employed by migrants as an alternative form to send remittances to their relatives in the homeland. Bitcoin has even been embraced by the Venezuelan government as a way to avoid international economic sanctions, and to recover control over its national economic systems. The South American government has endowed huge expectations on their new virtual coin called "The Petro," nonetheless as in the case of oil, the most important source of revenues in the country, this virtual currency has had a very turbulent road in its few years of life. But as it is possible to read in this chapter, no matter the obstacles, cryptocurrencies will endure and grow in Latin America, mainly because they have proven to be a more consistent form of payment to sell and acquire goods and services, particularly in countries in where the people have lost faith in their national currency and leaders.

Final remarks

As robots, digital applications, artificial intelligence systems, and drones become familiar in our daily lives, novel concerns arise. To take but one illustrative example, consider the case of the Uber autonomous car accident in Arizona in 2018. Who is legally liable? Is it the autonomous driver, the human driver, the company, or the city? What kinds of limitations should be made to the artificial intelligence's capacity for making decisions? These are no longer ethical and legal affairs that exist only in science fiction books and movies, but real problems that are shaping our daily coexistence with "thinking" machines. In order to provide clear legal frameworks and harmonious ethical principles to guide the arrival and adoption of emerging technologies, the involvement of government and civil society at a local and international level is mandatory. Biased visions could lead to unbending and prejudicial actions that will limit not only the development and global access to these emerging technologies but could also establish oppressive and even illegal practices. As complex artificial intelligence systems and robots are continuing to evolve across diverse economic sectors from medical services to security, there is a growing uncertainty concerning how to enhance traditional legal and moral authority to cover social relations with those "entities." One example is the question of how to deal with the fact that many cities' public services, like the water supply, traffic, and public parking will no longer be controlled by humans but by AI programs. In this context, some scholars have criticized the omnipresence of information technologies in urban settings, suggesting alternative expansion models more protective of the environment and less dependent on connectivity and technology.

Regarding Latin America, in spite of the significant economic and technological differences with developed settings like in Europe, China, or the United States, this new century has witnessed an impressive rise in the adoption level of digital technologies, from the Internet to smart phones and other personal devices. People spend most of their time on social media platforms interacting and sharing information among themselves. Even the presence of emerging technologies like drones, domestic robots, and machine learning applications are more frequent in diverse contexts of the Latin American daily life: in factories, at homes, in schools or used by the army or the police force, and even operated by crime organizations like Narco-gangs. The actual Latin American landscape is an intricate "cocktail" in which old fashioned and obsolete technologies like dial-up internet access converges with cutting-edge technology like robots and artificial intelligence in factories and institutes. The COVID-19 pandemic appeared amid a technological transition in Latin America, in which drones and robotic technology intermarry with very rudimentary communication media like civil band radio and obsolete personal desktops. This disparity will eventually have a huge impact not only regarding the casualties caused by the pandemic but the later economic recovery.

In this contradictory situation it is possible to find out-of-date polices, norms, and ethical regulations, dated even before the creation of the Internet, that still regulates people's activities online. It is precisely this amalgamation of paradoxical situations that encourages the necessity to undertake a more critical and deeper analysis into a series of cases and incidents across the region that have not been studied: like the use of drones in police surveillance, the invasion of privacy and the lack of proper regulation frameworks to protect users, and the concerns about the automation and the incursion of robots into the factories and the detriment and loss of labor rights. The works contained in this first volume stand as an important effort to fill, at least partly, this incipient reality that deserved a critical and integral approach.

References

Angelini, D., & Gibson, S. (2007). Organized Crime and Technology. *Journal of Security Education*, 2(4), 65–73. https://doi.org/10.1300/J460v02n04_07

Bailey, A. A., Bonifield, C. M., & Arias, A. (2018). Social media use by young Latin American consumers: An exploration. *Journal of Retailing and Consumer Services*, *43*, 10–19. https://doi.org/10.1016/j.jretconser.2018.02.003

Berg, A., Buffie, E. F., & Zanna, L.-F. (2016). Robots, Growth, and Inequality. *Finance & Development*, *53*(3). http://www.imf.org/external/pubs/ft/fandd/2016/09/berg.htm

Blankstein, A., & Romero, D. (2019, marzo 19). U.S. «virtual kidnapping» investigation focuses on Mexico. *NBC News*. https://www.nbcnews.com/news/crime-courts/virtual-kidnapping-calls-may-be-coming-mexico-u-s-officials-n985186

Cawley, M. (2017, March 27). Drone Use in Latin America: Dangers and Opportunities. *InSight Crime*. https://www.insightcrime.org/news/analysis/drone-use-in-latin-america-dangers-and-opportunities/

Ceurvels, M. (2019). *Latin America Digital Video 2019*. eMarketer. https://www.emarketer.com/content/latin-america-digital-video-2019

Correa, T. (2016). Digital skills and social media use: How Internet skills are related to different types of Facebook use among 'digital natives'. *Information, Communication & Society, 19*(8), 1095–1107. https://doi.org/10.1080/1369118X.2015.1084023

Dunn, D. H. (2013). Drones: Disembodied aerial warfare and the unarticulated threat. *International Affairs, 89*(5), 1237–1246. https://doi.org/10.1111/1468-2346.12069

EFE. (2020, mayo 14). Robots dan incansable ayuda contra el coronavirus. *El Informador.* https://www.informador.mx/robots_estirilizan-vf20200514mp4.html

Heinemeyer, M. (2019, junio 28). *How cyber-criminals are exploiting Latin America's new digital economy.* https://www.idgconnect.com/opinion/1502185/cyber-criminals-exploiting-latin-america-digital-economy

Human Rights Watch. (2020). *Mobile Location Data and Covid-19: Q&A.* https://www.hrw.org/news/2020/05/13/mobile-location-data-and-covid-19-qa

Keene, S. D. (2011). Emerging threats: Financial crime in the virtual world. *Journal of Money Laundering Control, 15*(1), 25–37. https://doi.org/10.1108/13685201211194718

Levin, A., & Beene, R. (2018, July 8). Attack on Venezuela's president highlights U.S. vulnerability to drone assassins. *Chicagotribune.Com.* https://www.chicagotribune.com/news/nationworld/politics/ct-venzuela-drone-security-laws-20180807-story.html

Plaw, A., Fricker, M. S., & Colon, C. (2015). *The Drone Debate: A Primer on the U.S. Use of Unmanned Aircraft Outside Conventional Battlefields.* Maryland: Rowman & Littlefield Publishers.

Rempfer, K. (2019, September 23). Information warfare should be treated like call-for-fire missions, Army Cyber says. *Army Times.* https://www.armytimes.com/news/your-army/2019/09/17/information-warfare-should-be-treated-like-call-for-fire-missions-army-cyber-says/

Ruiz-del-Solar, J., & Weitzenfeld, A. (2012). Advances in Robotics in Latin America. *Journal of Intelligent & Robotic Systems, 66*(1–2), 1–2. https://doi.org/10.1007/s10846-011-9629-6

Statista Research Department. (2020a). *Internet usage among children in Latin America 2019.* Statista. https://www.statista.com/statistics/1055900/internet-usage-children-latin-america/

Statista Research Department. (2020b). *Social media penetration in Latin America & the Caribbean 2020.* Statista. https://www.statista.com/statistics/454805/latam-social-media-reach-country/

Taylor, J. (2020, April 19). Facebook and Google to be forced to share advertising revenue with Australian media companies. *The Guardian.* https://www.theguardian.com/media/2020/apr/19/facebook-and-google-to-be-forced-to-share-advertising-revenue-with-australian-media-companies

The Economist. (2018, February 8). Home-made drones now threaten conventional armed forces. *The Economist.* https://www.economist.com/news/science-and-technology/21736498-their-small-size-and-large-numbers-can-overwhelm-defences-home-made-drones-now

Vosoughi, S., Roy, D., & Aral, S. (2018). The spread of true and false news online. *Science, 359*(6380), 1146–1151. https://doi.org/10.1126/science.aap9559

Yadron, D. (2016, February 2). Fisher-Price smart bear allowed hacking of children's biographical data. *The Guardian*. https://www.theguardian.com/technology/2016/feb/02/fisher-price-mattel-smart-toy-bear-data-hack-technology

Part I

Digital data and privacy, prospects, and barriers

2 The reception of sexual messages among young Chileans and Uruguayans

Predictive factors and perception of harm[1]

Amaranta Alfaro, Matías Dodel,
and Patricio Cabello

Introduction

From media advertising and magazine covers to song lyrics, posts, and memes, we receive and share, willingly or unwillingly, sexual content, and it has become a part of our everyday lives. As Burkett argues, we are immersed in an "increasingly sexualized culture" (2015 p.836). Whereas sex and sexual expression is – and always has been – inherently social (Burkett, 2015 p.842), social media and the new digital communication arena are reshaping its dynamics, particularly for young people (De Ridder, 2017; Jabaloyas, 2015). As with the massification of all new media technologies, new practices have arisen, as well as new moral panics concerning how these practices affect young people (Tatar, 1998). In the context of this chapter, the reception of sexual messages is considered as part of sexting practices. Sexting is a term that combines the words sex and texting (Livingstone & Görzig, 2012 p.151) that is mostly used by "experts" and "adults" in media and academic discourses (Burkett, 2015 p.843) and encompasses a set of diverse practices.

Sexting has been defined as the creating, sharing, and forwarding of sexually explicit and/or suggestive messages, and/or nude or nearly nude images or photos communicated through electronic means, primarily via smartphones, and web 2.0 activities such as social networking sites (Klettke et al., 2014 p.45; Lenhart, 2009; Ringrose et al., 2012 p.9; Weisskirch & Delevi, 2011). Sexting implies a variety of content such as asking for, sending, and receiving nude or semi-nude pictures, photos or videos of sexual acts, but also text messages with any kind of sexual connotation (Burkett, 2015; De Ridder, 2017; Dobson, 2018).

Taking into account that sexting has become not an uncommon behavior among teenagers, (Burkett, 2015 p.851; Quesada, Fernández-González & Calvete, 2018; Ringrose & Harvey, 2015; Van Ouytsel et al., 2017 p.461) there is a need to approach it without moral panic. Certain media and even academic approaches to the phenomenon describe it as risky, harmful, and deviant for young people (Burkett, 2015; Kosenko et al., 2017), labeling it "a new type of deviant sexualized behavior in youth that is associated with many risks" (Döring, 2014 p.4). As Haddon and Livingstone (2012) have argued, this is part of the pedagogic and parental discourses that focus more on risks rather than on the opportunities of the Internet. This chapter will study one sexting-associated practice: the reception of messages with sexual content among adolescents from two high-income Latin American countries, Chile and Uruguay. We will assess the factors predicting the reception of messages with sexual content by young people, signaling the commonalities and differences between significant predictors in these two countries, as well as studying the impact of receiving messages with this content for Chilean youth's perception of harm.

Regarding context, in Chile there is a lack of reliable information about sexting. Exploratory research shows that in schools (boys and girls from five to 18 years-old) who receive messages with sexual content have a prevalence between 5.8% (private schools) to 7.7% (public schools). On the one hand, sending these kind of messages is more frequent in private schools (6,1%) than public ones (5%). On the other hand, both receiving and sending this kind of messages is more prevalent between male children than female children (Arias, Buendía & Fernandez, 2018). It is important to highlight that Chile has one of the most segregated educational systems where public schools show significantly lower rates in quality indicators than private schools (Valenzuela, Bellei & Rios, 2014; Bellei, Contreras, Canales & Orellana, 2018), therefore these differences show again how inequalities reflect on digital media experiences. In regard to Chilean legal aspects, despite sexting being considered a relevant phenomenon according to the law, it is not in itself illegal and people, adults and children, that send personal sexual messages are not prosecuted by law. Instead, the publication or any form of dissemination of personal information of others, including nudity pictures and sexual messages without their consent, is considered a felony (Scheechler, 2019). While in Uruguay, where information on this subject is also scarce, according to data from the Ministry of Interior in 2013, 120 complaints were filed for sexting, sextortion, or grooming against girls, boys, and women, in one year – 2014 – this figure tripled to 450 cases with 19 people prosecuted, turning into 700 cases in 2016 (LaRed21, August 26, 2016; Subrayado, November 1st, 2018). Besides this data there is not much more information, since so far it has been a topic mainly studied with a focus on cyberviolence towards women (see Cestau, 2018).

Contrary to media assessment and public opinion concerning the subject (Burkett, 2015; Kosenko et al., 2017), sexting occurs both in sexual contexts (dating, casual, or established relationships), and non-sexual peer exchanges among friends (Burkett, 2015) with no intentional "sexual" nature (Albury, 2015) such as jokes (Burkett, 2015 p.859; Korenis & Billick, 2014; Lippman & Campbell, 2014; Yépez-Tito, et al., 2018). Some scholars have even stated that joke-related sexting tends to be most prevalent among youth (Symons et al., 2018). Whatever the case, this kind of online messaging can have unwanted consequences, to the point of becoming an unpleasant or problematic experience for some children and adolescents (Karrera & Garmendia, 2018).

We consider sexting as a sexual technology-mediated expression and experimentation practice, exercised by young people – beyond the control of adults (De Ridder, 2017 p.4) – given the digital context in which they are immersed (Dobson, 2018 p.93; Döring, 2014; Lenhart, 2009; Livingstone et al., 2011). This confronts them with the ambivalence of managing their own sexual agency, their own reputations, self-expression, developing an online identity, and reinforcing their own body image (Angelides, 2013; Bianchi et al., 2017; Walrave & Van Ouytsel, 2014), under the tension of "regulated-freedom" as coined by Bragg and Buckingham (2004 p.245).

Factors related to sexting and sexual content sharing

Sexting has largely been associated with older adolescents (Baumgartner et al., 2014; Dake et al., 2012; Ybarra & Mitchell, 2014). There is a higher frequency of sexting behaviors in teenagers over 16 years old, thus situating the greatest difference in the practice of sending images and messages of sexual content (Quesada et al., 2018). In terms of gender, cultural stereotypes, and constructions about femininity and masculinity (Davidson, 2015; West et al., 2014), these carry different expectations for girls and boys in the practice of sexting and its negotiation process in terms of the meanings, motivations, roles, and consequences attached to it (Davidson, 2015; Marrufo, 2012; Ringrose et al., 2013; Symons et al., 2018 p.3839). Sexting seems to be more prevalent among boys (Yépez-Tito et al., 2018). They are perceived as the initiators and the ones who ask for and receive sexual photos (Baumgartner et al., 2014; Gordon-Messer et al., 2013; Temple et al., 2012; Yépez-Tito et al., 2018), while the girls are the ones feeling pressured into sexting (Döring, 2014; Englander, 2012; Walker et al., 2013) and are blamed (Hasinoff, 2014) and negatively judged because of it – namely through "slut-shaming" (Ringrose & Harvey, 2015; Symons et al., 2018 p.3839; Van Ouytsel et al., 2017; Walrave et al., 2014).

In terms of risks and problematic behaviors, sexting has been associated with other risky practices such as drug or alcohol consumption (Benotsch et al., 2013; Temple et al., 2014), dangerous sexual conduct – like unprotected sex – (Rice et al., 2012; Temple et al., 2012), and troublesome use of technology (Delevi & Weisskirsch, 2013). It has also been related to different types of victimization, such as harassment and cyberbullying (Dake et al., 2012; Quesada et al., 2018; Wachs et al., 2015; Ybarra et al., 2007), depression (Temple et al., 2014), anxiety (Drouin & Landgraff, 2012), and dating violence (Morelli et al., 2016; Quesada et al., 2018). Finally, regarding platform and devices, the literature signals that owning a smartphone connected to the Internet is closely related to sexting practices (Ghorashi et al., 2019; Ringrose et al., 2013). Sexting practices take place essentially on applications such as Snapchat and WhatsApp (Ghorashi et al., 2019; Van Ouytsel et al., 2017), while, to a lesser extent, applications such as Viber and Instagram (Ghorashi et al., 2019), and Facebook (Jane, 2017; Van Ouytsel et al., 2017) are also mentioned in sexting-related studies.

Consequently, this chapter aims to explore the following research questions:

- RQ1: what is the prevalence of the reception of messages with sexual content among Chilean and Uruguayan adolescents who use the Internet?
- RQ2: which factors predict the occurrence of the reception of messages with sexual content among Chilean and Uruguayan adolescents who use the Internet?
- RQ3: what is the impact of the perception of harm in the reception of messages with sexual content in wellbeing perceptions among Chilean adolescents who use the Internet?

First, our analysis delves into the factors that predict the reception of digital messages with sexual content. Using sex as a proxy, we inquire if gender determines the reception and reaction to this kind of messages. We also want to study the effects of age on their reception, as well as their links with other digital and non-digital risks, cultural capital (head of household education), and the app ecology these children inhabit. This chapter will also explore the potential negative effects of receiving such messages. As described, the present chapter will tackle a specific type of sexting-related behavior, focusing on the reception of messages with sexual content using data from the Kids Online Chilean and Uruguayan surveys. This chapter concentrates on identifying the factors associated with the reception of sexting-type messages, while also exploring the relationship between the experience of receiving messages of sexual content and the emotional response to such messages. We focus this second part of the chapter on teenagers who could have been harmed by the reception of sexting messages. This last point particularly answers the call for research expressed by Burkett (2015) about the need to explore the feelings associated with sexting and by Willard (2010) who stated the need to know more about the consequences of sexting as a social, cultural, or digital practice.

Data and methods

The data presented in this chapter come from the Chilean and Uruguayan versions of the Global Kids Online (GKO). Both surveys were developed with the objective of characterizing the opportunities and risks present on the Internet for children and adolescents to inform policies based on the voices and experiences of the children themselves. Both Chile and Uruguay's GKO surveys were conducted through computer-assisted personal interviews (CAPI) at children's homes by specially trained interviewers. The questionnaires had both face-to-face interview components and, for sensitive questions such as those related to sexual or non-conformant behaviors, private self-reporting using digital devices. Both questionnaires consisted of a local adaptation of the GKO core questionnaire.

In both countries, the surveys consisted of representative national samples of children and teenagers who were Internet users and aged between nine and 17 years (Chile, n=1000; Uruguay, n=948). Along with each child, a parent or guardian was also interviewed. The fieldwork in Chile was conducted between August and November 2016, while in the case of Uruguay, the fieldwork was conducted between August and December 2017. In relation to the topic at hand, whereas both Uruguay and Chile's Kids Online surveys assessed having received or seeing digital messages containing sexual content, in Chile respondents were aged 11 years old or higher, while in Uruguay they were aged 13 years old or higher. For ethical reasons, in both countries' issues related to the exchange of messages and images with sexual content were not raised among boys and girls aged nine and ten years old.

To assess the reception of digital messages with sexual content, two logistic binary regressions were fitted: the first, to predict the reception of digital messages with sexual content (both in Uruguay and Chile) and the second to predict harm as a consequence of receiving these kind of messages (only in Chile). As this last estimation required dealing with the bias of having received a message with sexual content, we included in our model a selectivity correction term (SCT) following the adaptation by Bucheli and Porzecanski (2011) of Heckman (1979) and Buchinsky (1998). The implementation is further detailed in the section on the model.

Findings and analysis

Descriptive statistics

Chilean and Uruguayan young people are extensive Internet and, predominantly, smartphone users: a significant number use these technologies for sexual interactions. Following the Kids Online survey, in both countries young people were asked if they had received messages with sexual content. The question was as follows: "In the PAST YEAR, have you EVER RECEIVED any sexual messages? This could be words, pictures, or videos." Positive responses were coded as "1," negative as "0." This question was adapted from the GKO Survey (Global Kids Online, 2019).

Table 2.1 presents the percentages of children who expressed having received digital messages with sexual content in the past year. Age categories were split every two years to make data comparable between the two countries. Regarding the whole Chilean sample, 21% stated they had received such a message in the past year. While among pre-adolescents (between 11 and 12 years old), this only reached 10%, the percentage slightly increased up to 14% when assessing adolescents between 13 and 14 years old, reaching 31% among the 15–17-year old in the Chilean group. Nonetheless, close to one in ten Chilean kids did not know if they had received messages with sexual content or did not want to answer the question. In Uruguay a slightly higher percentage of adolescents between 13 and 14 years old stated they had

Table 2.1 Chilean and Uruguayan children who reported receiving or seeing a message with sexual content in the past year in percentages, by age and gender

	Chile			Uruguay		
			Doesn't Know /Doesn't			Doesn't Know /Doesn't
	Yes	No	answer	Yes	No	answer
	(in %)	*(in %)*	*(in %)*	*(in %)*	*(in %)*	*(in %)*
Total	21	69	10	25	74	1
Female	20	69	11	18	81	1
Male	21	69	10	33	67	0
11–12 years	10	85	5	-	-	-
13–14 years	14	73	13	23	77	0
15–17 years	31	57	11	28	70	1

received such messages than in Chile (23%), but among the oldest groups, the responses were very similar across the two countries (28%).

Regarding age and gender-based distribution of the received messages, in both territories the older the teenager, the higher the percentage who had received messages with sexual content. Moreover, there seemed to be a strong gendered effect in the reception in Uruguay, which was almost double for boys than for girls. These results echo similar studies in the field (Baumgartner et al., 2014; Garmendia et al., 2018, Yépez-Tito et al., 2018). In the case of the Chilean study, the assessment of the reception of messages with sexual content went one step further, inquiring children how they felt after receiving or seeing those messages. The response categories were "very good," "good," "not good nor upset," "upset," and "very upset." This follow-up question analysis enabled the identification of a particular group of children that can be characterized as victimized or harmed because of receiving such messages. Consequently, a new binary variable was created, assuming a value of "1" if respondents felt "upset" or "very upset," and a "0" if they felt "very good," "good," or "neither good nor upset." In this sense, more than one quarter of the young people who received a message with sexual content (27%) felt upset or really upset, that is, 4% of all Chilean children (see Table 2.2).

Although comprising a smaller percentage of all Chilean youth compared to the receivers of the messages, victimized young people present a clear and distinct pattern of vulnerability regarding sex and age. In this sense, girls experienced quadruple the rate of boys in terms of harm, and younger pre-adolescents – between 11 and 12 years old – reported more than twice as much that they felt upset or very upset than the group between the ages of 13 and 17 years old. These results are in accord with similar studies, which showed that girls are more likely to feel upset about receiving messages or images of sexual content than boys (Garmendia et al., 2018).

Table 2.2 Chilean children who reported feeling upset or very upset after receiving or seeing a message with sexual content in the past year in percentages, by age and gender

	Prevalence of children who received or saw a message with sexual content in the past year and felt upset or very upset about it (n=762)	*Children who received or saw a message with sexual content in the past year and felt upset or very upset about it (n=160)*
Total (in %)	4	27
Female	7	46
Male	2	10
11–12 years	6	59
13–14 years	4	25
15–17 years	7	22

Predicting the reception of digital messages with sexual content

To predict the reception of digital messages with sexual content in both countries, identical logistic regressions were fitted based on Chile and Uruguay's Kids Online studies for teenagers aged 13 years and older only. Results, showed in table 2.3, are presented in odd ratios to ease the interpretation of the coefficients. The results show that receiving messages with sexual content shared a core of determinants in both countries, as predicted by the literature, such as gendered ratios – boys receive more of such messages than girls – (Baumgartner et al., 2014; Gordon-Messer et al., 2013; Temple et al., 2012; Yépez-Tito et al., 2018) and age-dependent reception – that is, it is more likely to occur in older adolescents (Baumgartner et al., 2014; Dake et al., 2012; Temple et al., 2012; Ybarra & Mitchell, 2014). Nonetheless, both Chile and Uruguay present differences in other statistically significant predictors.

Regarding the commonalities, gender – being male – was the strongest predictor of receiving such messages in Uruguay and the second one in Chile. Being a boy, compared to a girl, increased the odds of receiving or seeing these types of messages 4.28 times in Uruguay and 3.69 times in Chile (both at P<0.00). Cultural stereotypes of masculinity could be behind these striking odd disparities. Children who see sensitive or risky content online tended to receive more messages in both countries: each type of content increased the chances of receiving a message by 47% in Chile and 26% in Uruguay (P<0.00). Being a cyberbullying victim also statistically significantly increased the odds of receiving digital messages with sexual content compared to non-victims, 3.81 times in Chile (P<0.00), but only 1.42 times in Uruguay (P<0.05).

Finally, both models were also similar for several variables that are not statistically significant as predictors of the dependent variable – most surprisingly, the human capital of the parent (education) and the digital human capital of the kid (digital skills). Additionally, the app ecosystem in which the young people reside was not statistically significant at p<0.05 in either country. Nonetheless, the relevance of specific apps cannot be completely

Table 2.3 Predicting the reception of a message with sexual content in the past year in Chile and Uruguay (13 years and older), in odds ratios

	Uruguay		Chile	
	Odds Ratio	P>z	Odds Ratio	P>z
Male	4.28	0	3.69	0
Age	0.98	0.83	1.3	0.03
Education (middle)	1.66	0.19	0.57	0.11
Education (high)	1.93	0.15	0.59	0.22
Digital skills	0.94	0.09	1.03	0.23
Use Internet on Cellphone (all the time)	1.22	0.49	4.56	0
WhatsApp	2.51	0.07	0.57	0.45
Facebook	0.5	0.1	3.29	0.09
Twitter	0.92	0.8	0.89	0.76
Instagram	2	0.08	0.75	0.41
Saw online risk content (index)	1.26	0	1.47	0
Saw something only that made feel upset	1.45	0.27	2.6	0
Problems due to extensive use of the Internet	1.03	0.91	3.09	0
Problematic Offline behavior (2 or more)	2.04	0.03	1.83	0.11
Cyberbullying victims	1.42	0.05	3.81	0
Constant	0.11	0.25	0	0

McFadden's Adj R2 - URUGUAY: 0.197; Chile: 0.259

disregarded, as some of them are close to this threshold, such as WhatsApp and Instagram for the Uruguayan case.

Among divergent predictors, the highest frequency of Internet use through mobile phone (all the time), for example, was the strongest predictor of receiving sexual messages in Chile, increasing its chances by 356%, but had no statistically significant effect for Uruguayan youth. A linked construct, having have had problems due to extensive use of the Internet, also increased the odds of receiving such messages in Chile by 3.09 times (P<0.00). Similarly, whereas in Chile the older the respondent, the higher the odds of receiving messages with sexual content (OR=1.30, P<0.00), age had no significant effect in the Uruguayan case. Having seen something that made youth feel upset also was significant as a predictor for the Chilean case (OR=2.60, P<0.00), but not in Uruguay. On the other hand, Uruguayan youth with two or more offline problematic behaviors had a 104% greater chance of receiving a message with sexual content (P<0.03), than the ones with fewer offline problematic behaviors, something that was not statistically significant in Chile.

Predicting harm as a consequence of the reception of messages with sexual content

The second model, a logistic regression, estimated the predictors of harm as a consequence of the reception of messages with sexual content. Whereas, a logistic regression presented in table 2.4. We also fit a logistic binary model

Table 2.4 Predicting the harm due to the reception of a message with sexual content in the past year in Chile (11 years and older), in odds ratios with a selectivity correction term

	OR	P>z
11 & 12 years old	4.42	0.004
Male	0.17	0
Digital Skills	0.95	0.152
SCT (selectivity correction term)	0.21	0
Constant	2.78	0.42

(in this case for the whole inquired population in Chile, aged 11 and older), this estimation requires dealing with a truncated dependent variable or the bias of having received a message with sexual content. To correct for this bias, we replicated the strategy of Bucheli and Porzecanski (2011), based on the proposal of Heckman (1979) and developed initially by Buchinsky (1998). Like these authors, we first estimated a probit selection model replicating the logistic regression for the Chilean case in the first model (but including 11 and 12-year-old children, too), to then use the predicted probability to estimate the inverse of the Mills ratio to finally create an SCT. We then introduced the SCT in the logistic regression predicting damage as a consequence of receiving messages with sexual content. As there is no clear guide for how this term should be interpreted (Bucheli & Porzecanski, 2011), we only use the term to correct for selection bias.

Confirming the literature predictions, when receiving a message with sexual content, girls and younger adolescents expressed substantially higher levels of harm than boys and older adolescents (Livingstone & Görzig, 2014; Ringrose & Harvey, 2015; Walrave et al., 2015). Being 11 or 12 years old increased the chances of harm by 342% compared to being 13 years or older, and being a boy reduced by 83% these chances compared to being a girl. On one hand, the level of skills seemed to play no role here, just as in the case of receiving the message itself.

Conclusions

Based on our findings, receiving messages with sexual content is not an uncommon behavior among Chilean and Uruguayan adolescents, with almost one fourth of the youngsters stating they have received messages with sexual content (21% for Chile and 25% for Uruguay). Coincidentally with previous sexting studies, age and gender have a close relationship with the practice of receiving sexual messages, the reception of these messages being higher, the older the subject, and led by boys – particularly in Uruguay – over girls. This effect might relate to the role played by boys in the sexting behavior observed by precursory research, where they are usually perceived as the promoters and the ones who ask for sexual photos and accumulate them as a sort of trophy, because they enhance their reputations by bragging about such images among peers (Baumgartner et al., 2014; Gordon-Messer et al., 2013; Temple et al., 2012; Yépez-Tito et al., 2018). The above discussed cultural stereotypes of masculinity (West et al., 2014) is related to misogynistic joking or the showing off practices of boys' collections of sexual content as "proof" of their skills to chat with girls and "negotiate access to seeing their bodies" (Ringrose & Harvey, 2015; Ringrose et al., 2013 p.9; Yépez-Tito et al., 2018) and could explain these results for boys. It is worth noting that it is not advisable to demonize individualized young male behaviors, as their practices are immersed in a performative digital peer pressure context (Kofoed & Ringrose, 2012), which is a consequence of societal and cultural norms and values.

While seeing risky or sensitive content online, as well as being a victim of cyberbullying, increases the odds of receiving a message with sexual content in both countries, the app ecosystem that children inhabit plays no role. Nevertheless, some apps such as WhatsApp and Instagram – originally for smartphones – are relevant for the Uruguayan case, while in the case of Chile, the high frequency of Internet use through cellphones is a strong predictor for sexting. Both have in common, mobile phone usage – a personal device and away from parental control – that gives the young people the freedom to use and explore technology-mediated sexual expressions (De Ridder, 2017 p.4), echoing the claims of Ghorashi and colleagues (2019).

Additionally, our analysis validates the revised literature such that girls and younger adolescents express substantially higher levels of harm when receiving a message with sexual content (Garmendia et al., 2018; Livingstone & Görzig, 2014; Ringrose & Harvey, 2015; Walrave et al., 2015). This echoes Willard (2011) and Peterson-Iyer (2013), who describe the unsolicited or unwanted distribution of messages with sexual content as an act of harassment and high-tech bullying, which naturally implies that its reception could cause harm.

In the case of girls, this could go a step further, if one considers the misogynistic nature of the distributed content – memes and other messages of sexual character – also stated in literature as cyber gender-based harassment (Krieger, 2017). Using the sexting literature, one could understand some examples of the kind of harm perceived by girls in the reception of messages with sexual content, because they are exposed to other vulnerabilities, such as being targets of criticism and judgment – slut-shaming and revenge porn – when engaging in sexting practices (Döring, 2014; Ringrose et al., 2013; Symons et al., 2018 p.3839; Van Ouytsel et al., 2017). The self-guilt and shame created by these situations, perpetuates the victim-blaming dynamics according to the previous mentioned cultural frames and stereotypes; these cannot be assumed for the present data, but certainly present an example.

The present results point out the relevance of addressing sexting practices in the frame of sexual education and public health initiatives targeted to teenagers to promote a reflective and responsible attitude among adolescents.

These initiatives should begin, ideally, from an early stage and address the gender perspective, where higher levels of harm are perceived as consequences of sexting practices, such as receiving a message with sexual content.

Limitations and future work

Although we inquired only about receiving digital messages with sexual content and the measures used allowed a general overview of the reception of these messages, little can be inferred about the nature and content of those messages. In further research it would be relevant to address the characteristics of these risks through qualitative studies, following the strategies of recent studies such as that by Alonso-Ruido and colleagues (2018). An additional limitation of our study is that our data does not enable us to go beyond the reception of sexual messages and capture richer practices, such as the exchange of sexual messages that would attend to a more relational phenomenon. Finally, the self-reported nature of the survey and the sensitive character of the subject could imply a distortion in the responses. Cultural differences between these two countries must also be considered.

Regarding lines for future work, as we know that sending and receiving sexual messages is a phenomenon with low prevalence in the national sample, it is not suitable for more complex statistical modeling, and future research could overcome this limitation by using larger, school-based samples. Second, the exchange of messages, sexual or not, is less a matter of sending and receiving and more a matter of meaning. In that sense, we should ask children what it means for them to exchange this kind of information. How does this exchange take part in their construction of social relationships in a digital era? We believe that this kind of question should be explored more deeply through qualitative research, especially in developing countries like Chile and Uruguay.

Note

1. The development of this chapter was supported by Ministry of Education (Chile), the Regional Bureau for Education in Latin America and the Caribbean, UNESCO, and by Fondo de Investigación Fundamental Clemente Estable from Agencia Nacional de Investigación e Innovación (Uruguay), project FCE_3_2018_1_149415.

References

Albury, K. (2015). Selfies, sexts and sneaky hats: Young people's understandings of gendered practices of self-representation. *International Journal of Communication*, *9*, 12.

Alonso-Ruido, P., Rodríguez-Castro, & Lameiras-Fernández, M., & Martínez-Román, R. (2018). El Sexting a través del discurso de adolescentes españoles. *Saúde e Sociedade, 27*, 398–409.

Angelides, S. (2013). 'Technology, hormones, and stupidity': The affective politics of teenage sexting. *Sexualities, 16*(5–6), 665–689.

Arias Cerón, M., Buendía Eisman, L., & Fernández Palomares, F. (2018). Grooming, Ciberbullying y Sexting en estudiantes en Chile según sexo y tipo de administración escolar. *Revista chilena de pediatría*, 89(3), 352–360.

Baumgartner, S. E., Sumter, S. R., Peter, J., Valkenburg, P. M., & Livingstone, S. (2014). Does country context matter? Investigating the predictors of teen sexting across Europe. *Computers in Human Behavior, 34*, 157–164.

Bellei, C., Contreras, M., Canales, M., & Orellana, V. (2018). The Production of Socio-economic Segregation in Chilean Education: School Choice, Social Class and Market Dynamics. *Understanding School Segregation: Patterns, Causes and Consequences of Spatial Inequalities in Education*, 221–242.

Benotsch, E. G., Snipes, D. J., Martin, A. M., & Bull, S. S. (2013). Sexting, substance use, and sexual risk behavior in young adults. *Journal of Adolescent Health, 52*(3), 307–313.

Bianchi, D., Morelli, M., Baiocco, R., & Chirumbolo, A. (2017). Sexting as the mirror on the wall: body-esteem attribution, media models, and objectified-body consciousness. *Journal of Adolescence, 61*, 164–172.

Bragg, S., & Buckingham, D. (2004). *Young people, sex and the media*. New York, NY: Palgrave Macmillan.

Bucheli, M., & Porzecanski, R. (2011). Racial inequality in the Uruguayan labor market: An analysis of wage differentials between Afro-descendants and whites. *Latin American Politics and Society, 53*(2), 113–150.

Buchinsky, M. (1998). The dynamics of changes in the female wage distribution in the USA: A quantiles regression approach. *Journal of Applied Econometrics*, 13, 1–30.

Burkett, M. (2015). Sex(t) talk: A qualitative analysis of young adults' negotiations of the pleasures and perils of sexting. *Sexuality & Culture, 19*(4), 835–863.

Cestau, I. (2018). Cuando lo virtual es real: ciberviolencia contra las mujeres. In FLACSO & CES "Hacia vínculos afectivos libres de violencia Aportes para el abordaje educativo de jóvenes y adolescentes. Tomo II". ANEP: Montevideo Available online from: https://www.ces.edu.uy/files/2018/+CES/ProCI/Hacia_vinculos_afectivos_libres_de_violencia_-_TOMO_II.pdf

Dake, J. A., Price, J. H., Maziarz, L., & Ward, B. (2012). Prevalence and correlates of sexting behavior in adolescents. *American Journal of Sexuality Education, 7*(1), 1–15.

Davidson, J. (2015). *Sexting: Gender and teens*. Boston, MA: Sense Publishers.

Delevi, R., & Weisskirch, R. S. (2013). Personality factors as predictors of sexting. *Computers in Human Behavior, 29*(6), 2589–2594.

De Ridder, S. (2017). Social media and young people's sexualities: Values, norms, and battlegrounds. *Social Media + Society, 3*(4), 2056305117738992. https://doi.org/10.1177/2056305117738992

Dobson, A. S. (2018). Sexting, intimate and sexual media practices, and social justice. In A. S. Dobson, B. Robards, & N. Carah (Eds.) *Digital Intimate Publics and Social Media* (pp. 93–110). Cham, Switzerland: Palgrave Macmillan.

Döring, N. (2014). Consensual sexting among adolescents: Risk prevention through abstinence education or safer sexting? *Cyberpsychology: Journal of Psychosocial Research on Cyberspace, 8*(1). http://dx.doi.org/10.5817/CP2014-1-9.

Drouin, M., & Landgraff, C. (2012). Texting, sexting, and attachment in college students' romantic relationships. *Computers in Human Behavior, 28*(2), 444–449.

Englander, E. (2012). *Low risk associated with most teenage sexting: A study of 617 18-year-olds*. Bridgewater, MA: Bridgewater State University.

Garmendia, M., Casado, M.A., Jiménez, E., & Garitaonandia, C. (2018) Oportunidades, riesgos, daño y habilidades digitales de los menores españoles. In E. Jiménez, M. Garmendía, & M. Á. Casado (Eds.). *Entre selfies y whatsapps. Oportunidades y riesgos para la infancia y la adolescencia conectada* (pp. 31–42). Barcelona: Gedisa.

Global Kids Online (2019). *Global Kids Online Survey*. London: LSE-Global Kids Online.

Ghorashi, Z., Loripoor, M., & Lotfipur-Rafsanjani, S. M. (2019). Mobile access and sexting prevalence in high school students in Rafsanjan City, Iran in 2015. *Iranian Journal of Psychiatry and Clinical Psychology*, *24*(4), 416–425.

Gordon-Messer, D., Bauermeister, J. A., Grodzinski, A., & Zimmerman, M. (2013). Sexting among young adults. *Journal of Adolescent Health*, *52*(3), 301–306.

Haddon, L., & Livingstone, S. (2012). EU Kids Online: National perspectives. URL: http://www.lse.ac.uk/media@lse/research/EUKidsOnline/EU%20Kids%20III/Reports/PerspectivesReport.pdf

Hasinoff, A. A. (2014). Blaming sexualization for sexting. *Girlhood Studies*, *7*(1), 102–120.

Heckman, J. J. (1979). Sample selection bias as a specification error. *Econometrica*, *47*(1), 53–162.

Jabaloyas, C. (2015). Las TICs como factor de riesgo de la violencia en parejas adolescentes. *Criminología y Sociedad*, *4*, 211–264.

Jane, E. A. (2017). Feminist Digilante responses to a slut-shaming on Facebook. *Social Media + Society*, *3*(2), 2056305117705996. https://doi.org/10.1177/2056305117705996

Karrera, I. & Garmendia, M. (2018) Sexting: Qué sabemos y qué nos queda por aprender. In E. Jiménez, M. Garmendía, & M. Á. Casado (Eds.). *Entre selfies y whatsapps. Oportunidades y riesgos para la infancia y la adolescencia conectada* (pp. 43–54). Barcelona: Gedisa.

Klettke, B., Hallford, D. J., & Mellor, D. J. (2014). Sexting prevalence and correlates: A systematic literature review. *Clinical Psychology Review*, *34*(1), 44–53.

Kofoed, J., & Ringrose, J. (2012). Travelling and sticky affects: Exploring teens and sexualized cyberbullying through a Butlerian-Deleuzian-Guattarian lens. Discourse: Studies in the Cultural Politics of Education, 33(1), 5–20.

Korenis, P., & Billick, S. B. (2014). Forensic implications: Adolescent sexting and cyberbullying. *Psychiatric Quarterly*, *85*(1), 97–101.

Kosenko, K., Luurs, G., & Binder, A. R. (2017). Sexting and sexual behavior, 2011–2015: A critical review and meta-analysis of a growing literature. *Journal of Computer-Mediated Communication*, *22*(3), 141–160.

Krieger, M. A. (2017). Unpacking "sexting": A systematic review of nonconsensual sexting in legal, educational, and psychological literatures. *Trauma, Violence, & Abuse*, *18*(5), 593–601.

LaRed21 (2016, August 26). Ministerio del Interior recibió 700 denuncias de pornografía en línea por delitos de sexting, sextorsión y grooming en 2016. Retrieved from http://www.lr21.com.uy/comunidad/1302163-700-denuncias-pornografia-internet-sexting-sextorsion-grooming

Lenhart, A. (2009). Teens and sexting: How and why minor teens are sending sexually suggestive nude or nearly nude images via text messaging. Pew Research Centre Report. *Pew Internet & American Life Project, 1*, 1–26.

Lippman, J. R., & Campbell, S. W. (2014). Damned if you do, damned if you don't... if you're a girl: Relational and normative contexts of adolescent sexting in the United States. *Journal of Children and Media, 8*(4), 371–386.

Livingstone, S., & Görzig, A. (2012) Sexting. In: S. Livingstone, L. Haddon, and A. Görzig (Eds.) *Children, risk and safety on the Internet: Research and policy challenges in comparative perspective* (pp. 151–164). Bristol: The Policy Press.

Livingstone, S., & Görzig, A. (2014). When adolescents receive sexual messages on the Internet: Explaining experiences of risk and harm. *Computers in Human Behavior, 33*, 8–15.

Livingstone, S., Haddon, L., Görzig, A., & Ólafsson, K. (2011). *Risks and safety on the Internet: The perspective of European children*. Full findings, London: LSE EU Kids Online.

Marrufo, M. R. O. (2012). *Surgimiento y proliferación del sexting. Probables causas y consecuencias en adolescentes de secundaria*. Master's thesis. Investigación Educativa, Universidad Autónoma de Yucatán, México.

Morelli, M., Bianchi, D., Baiocco, R., Pezzuti, L., & Chirumbolo, A. (2016). Sexting, psychological distress and dating violence among adolescents and young adults. *Psicothema, 28*(2), 137–142.

Peterson-Iyer, K. (2013). Mobile porn? Teenage sexting and justice for women. *Journal of the Society of Christian Ethics, 33*, 93–110. doi:10.1353/sce.2013.0036

Quesada, S., Fernandez-Gonzalez, L., & Calvete, E. (2018). El sexteo (sexting) en la adolescencia: Frecuencia y asociación con la victimización de ciberacoso y violencia en el noviazgo. *Behavioral Psychology/Psicologia Conductual, 26*(2), 225–242.

Rice, E., Rhoades, H., Winetrobe, H., Sanchez, M., Montoya, J., Plant, A., et al. (2012). Sexually explicit cell phone messaging associated with sexual risk among adolescents. *Pediatrics, 130*(4), 667–673.

Ringrose, J., Gill, R., Livingstone, S., & Harvey, L. (2012). A qualitative study of children, young people and 'sexting': A report prepared for the National Society for the Prevention of Cruelty to Children. London.

Ringrose, J., & Harvey, L. (2015). Boobs, back-off, six packs and bits: Mediated body parts, gendered reward, and sexual shame in teens' sexting images. *Continuum, 29*(2), 205–217.

Ringrose, J., Harvey, L., Gill, R., & Livingstone, S. (2013). Teen girls, sexual double standards and 'sexting': Gendered value in digital image exchange. *Feminist Theory, 14*(3), 305–323.

Scheechler, C. (2019). Aspectos fenomenológicos y políticos-criminales del sexting. Aproximación a su tratamiento a la luz del Código Penal chileno. *Política criminal, 14*(27), 376–418.

Subrayado (2018, November 1). Secundaria advierte qué hacer a jóvenes que sean víctimas de "pornovenganza". Retrieved from https://www.subrayado.com.uy/secundaria-advierte-que-hacer-jovenes-que-sean-victimas-pornovenganza-n517894

Symons, K., Ponnet, K., Walrave, M., & Heirman, W. (2018). Sexting scripts in adolescent relationships: Is sexting becoming the norm? *New Media & Society, 20*(10), 3836–3857.

Tatar, M. (1998). "Violent delights" in children's literature. In. J. Goldstein (Ed.) *Why we watch: The attractions of violent entertainment* (pp. 69–87). New York, NY: Oxford University Press.

Temple, J. R., Paul, J. A., Van Den Berg, P., Le, V. D., McElhany, A., & Temple, B. W. (2012). Teen sexting and its association with sexual behaviors. *Archives of Pediatrics & Adolescent Medicine, 166*(9), 828–833.

Temple, J. R., Le, V. D., van den Berg, P., Ling, Y., Paul, J. A., & Temple, B. W. (2014). Brief report: Teen sexting and psychosocial health. *Journal of Adolescence, 37*(1), 33–36.

Valenzuela, J. P., Bellei, C., & Ríos, D. D. L. (2014). Socioeconomic school segregation in a market-oriented educational system. The case of Chile. *Journal of education Policy*, 29(2), 217–241.

Van Ouytsel, J., Van Gool, E., Walrave, M., Ponnet, K., & Peeters, E. (2017). Sexting: Adolescents' perceptions of the applications used for, motives for, and consequences of sexting. *Journal of Youth Studies, 20*(4), 446–470.

Wachs, S., Junger, M., & Sittichai, R. (2015). Traditional, cyber and combined bullying roles: Differences in risky online and offline activities. *Societies, 5*(1), 109–135.

Walker, S., Sanci, L., & Temple-Smith, M. (2013). Sexting: Young women's and men's views on its nature and origins. *Journal of Adolescent Health, 52*(6), 697–701.

Walrave, M., Heirman, W., & Hallam, L. (2014). Under pressure to sext? Applying the theory of planned behaviour to adolescent sexting. *Behaviour & Information Technology, 33*(1), 86–98.

Walrave, M., Ponnet, K., Van Ouytsel, J., Van Gool, E., Heirman, W., & Verbeek, A. (2015). Whether or not to engage in sexting: Explaining adolescent sexting behaviour by applying the prototype willingness model. *Telematics and Informatics, 32*(4), 796–808.

Walrave, M., & Van Ouytsel, J. (2014). *Mediawijs online: jongeren en sociale media.* Amsterdam: LannooMeulenhoff.

West, J. H., Lister, C. E., Hall, P. C., Crookston, B. T., Snow, P. R., Zvietcovich, M. E., & West, R. P. (2014). Sexting among Peruvian adolescents. *BMC Public Health, 14*(1), 811.

Weisskirch, R. S., & Delevi, R. (2011). "Sexting" and adult romantic attachment. *Computers in Human Behavior, 27*(5), 1697–1701.

Willard, N. (2010). Sexting and youth: Achieving a rational response. *Journal of Social Sciences, 6*(4), 542–562.

Willard, N. E. (2011). School response to cyberbullying and sexting: The legal challenges. *Brigham Young University Education and Law Journal, 1*, 75–125.

Ybarra, M. L., & Mitchell, K. J. (2014). "Sexting" and its relation to sexual activity and sexual risk behavior in a national survey of adolescents. *Journal of Adolescent Health, 55*(6), 757–764.

Ybarra, M. L., Mitchell, K. J., Finkelhor, D., & Wolak, J. (2007). Internet prevention messages: Targeting the right online behaviors. *Archives of Pediatrics & Adolescent Medicine, 161*(2), 138–145.

Yépez-Tito, P., Ferragut, M., & Blanca, M. J. (2018). Prevalence and profile of sexting among adolescents in Ecuador. *Journal of Youth Studies, 22*(4), 505–519.

3 Small Data, Big Data, and the ethical challenges for a fragmented developing world

Peru's need for diversity-aware public policies on information technologies and practices

Hugo Claros

Introduction

The contemporary global dynamic can be characterized by the extreme importance of information. Information flows are increasingly frequent, diverse, and intense. These are possible thanks to the continuous evolution of technological elements that multiply, diversify, and decentralize the opportunities to produce and use information. "Big Data" refers to the availability of large, varied, and frequently updated datasets, as famously defined by Laney (2001). Extracting value from said data is possible thanks to the availability of bigger computational power and adequate practices and technologies to store it, retrieve it, and process it, generating more frequent, detailed, and interconnected insights. In that context, the question of how much Big Data may contribute to bridging living standards gaps in developing countries arises, both between countries and within them. The gap can be dramatic in key topics and having better information creates the opportunity to deploy more precise actions. This chapter presents related academic literature, reviewing some of the main risks, challenges, possibilities, and ethical dilemmas, and ponders their implications in Peru's case as a heterogeneous national context in which it is necessary to democratize the possibilities of extracting value from data, while promoting a safety-first approach that protects citizens from undesirable practices.

The proper production, handling, and use of data require experience, this experience conditions the quality and range of the information and, therefore, of the derived decision-making. Data is a resource. The insights produced by its analysis guide actors and contribute to either consolidation or erosion of their position. The necessary knowledge to produce and benefit from information is asymmetrically distributed throughout society. The ability to extract value from data in a more successful and efficient manner requires not only technical know-how, but also strategic judgment that allows to prioritize specific needs and assign available processing capability. Even if Big

Data makes it possible to obtain new answers to solving old problems, the potential availability of these Big Data sets does not replace nor necessarily surpass the possibilities afforded by less complex datasets. Likewise, many of the problems and deficiencies associated with the current handling of those traditional datasets are amplified when dealing with Big Data and its associated technologies. In Peru's case, its historical specificity is characterized by the need to increase both the availability of data and the population data awareness and literacy in order to tackle old and new challenges.

General risks and challenges associated with Big Data

Given the importance of the changes generated by Big Data, it is appropriate to present some of the main challenges and risks related to its adoption.

New technical possibilities for an asymmetrical world

A systematic analysis of the academic literature about Big Data and the ethical concerns raised in biomedical contexts (Mittelstadt & Floridi, 2016) identified five main topics: informed consent, privacy, ownership, epistemology, and the "Big Data divide." Through informed consent, people authorize third parties to generate data about them and use it with diverse objectives. With Big Data, consent becomes more challenging and nebulous due to the massiveness and reusability of datasets. This increases the need for a regulatory legislation that establishes with greater clarity the obligations surrounding the handling of third-party data. Furthermore, issues about informed consent reach a new level of difficulty considering that it is possible to reverse engineer and deanonymize data sets, which circumvents traditional consent mechanisms.

Information protection is also crucial due to the ever-increasing availability of sensitive information about people and the constant development of new techniques to get access to this information. There is research that shows the possibility of using online advertising packages to generate data about a specific user's location, making it possible to deploy individual third-party low-budget surveillance (Vines, Roesner, & Kohno, 2017). It is also possible to reverse the anonymization of spatial data: "Four randomly chosen points are enough to uniquely characterize 95% of the users (e > .95), whereas two randomly chosen points still uniquely characterize more than 50% of the users (e > .5). This shows that mobility traces are highly unique, and can, therefore, be re-identified using little outside information" (de Montjoye, Hidalgo, Verleysen, & Blondel, 2013, p. 2).

Likewise, it is not only individual information but group information that can be reverse engineered:

> Some authors see the fact that central to big data processes are no longer personal identifying information as a focal point, but rather

group information or data about units or categories. This is why it might be useful to broaden the term 'personal data' so that it encompasses not only data about natural persons or about legal persons, but also about groups ... In addition, it is important to point out that even although data may be aggregated, anonymized or encrypted, reversing de-identification is increasingly easy in the big data era. That is why several authors have suggested that it is important to critically assess aggregated and anonymized data, that is, to regulate group profiles and statistical data alongside personal data. (Taylor, 2016, p. 231)

Finally, the "Big Data divide" is created by the asymmetry between individuals and organizations: individuals as owners of the raw material which these big volume datasets are built on, and the organizations as owners of platforms to analyze and benefit from that data.

One of the main responses to these issues is the attempt to establish ethical principles about Big Data, which would imply: "at the highest level ... such things as integrity, honesty, objectivity, responsibility, trustworthiness, impartiality, non-discrimination, transparency, accountability, fairness, robustness, resilience, usability, efficiency, and independence" (Hand, 2018). However, the diversity of actors who adopt those principles creates concern about potential bad practices, like the collection of individual data supposedly in the interest of social betterment, when in reality the production of such datasets benefits some specific private interest, like the increase of unequal wealth distribution, the increase of social conflict for political benefit, the increase of people's dependence on technology, the undermining of individual integrity, etc. (Jurkiewicz, 2018). For a specific example about the privacy-related risks in healthcare, Terry (2014) points out that:

In parallel to the data-led transformation of the healthcare system itself, there is an increased demand for medical data outside traditional health care or the traditional health-care relationship. Unfortunately, some of this demand is being satisfied by criminal activity. Medical data theft has been steadily increasing, with the healthcare sector becoming the leading target for cyberattacks in 2013. The largest customers for legally acquired medical data are data brokers who sell business intelligence based on their processing of "big data." These data brokers are collecting, storing, and analyzing petabytes of medical data. The sources of these data are extremely varied but include medically inflected data derived, for example, from social media trails or over-the-counter retail. (p.836)

An additional concern involves the difference between data ownership and effective use of data, since in daily practice data may be used by any member of an organization, making data an enterprise-wide object, even if the formal use of said data inside an organization is limited to some specific team

(Zwiegelaar & Stylos, 2018). Other authors have identified the possibility to establish personalized prices according to each consumer's characteristics as a Big Data-related risk (Steinberg, 2019). There exists a risk of distorting the process of granting and obtaining consent from the general public (Lipworth, Mason, Kerridge, & Ioannidis, 2017), and the reality that the rising complexity of analysis techniques may limit the possibility that users can understand and interpret findings made based on their information (Lipworth et al., 2017). Given these multiple risks, it is important to reflect and be vigilant about how data is being produced and used in society.

The epistemological risk of overlooking the historical specificity of Big Data adoption in the developing world

The adoption of Big Data production and handling practices will not be similar to, nor will it represent what it did for developed countries in their day. There are two main reasons. Firstly, the adoption of Big Data as a new technology in developing countries takes place in a very different context from the one faced in the global North: having a new technology of massive data generation, with near real-time updating and processing, in a population with generalized coverage of minimum living standards, is not the same as when it emerges among highly heterogeneous population that do not have generalized basic services coverage. Furthermore, the relative position occupied by nation-states that comprise the global South is very different from the position which the global North occupied when Big Data emerged as a new technology.

While Big Data was initially a measure of innovation and competence between developed countries, a sign of global leadership; in developing countries, the progressive adoption of Big Data was from the beginning an exercise in adaptation to a trend with an agenda which is heavily influenced by the problems considered relevant to developed countries. Big Data emerges in developed countries, in the middle of the specific problems and challenges of those societies, without considering the specific problems of different ones, like those of developing countries. Because of this discrepancy, even if developing countries were to apply the same recipe that developed countries used in their attempts to tackle Big Data and its associated technologies, it would not result in a functionally similar result in terms of what this historical moment represented for the global North.

The incorporation of Big Data as a new technology in developing countries represents, for their societies, the challenge of integrating elements from the XXI century in a context that is still tackling problems solved by developed countries at the start of the XX century. Thus, in order for the adoption of Big Data and associated technologies to gain momentum towards the improvement of developing countries, it will be necessary that new strategies are generated, creatively combining the lessons and best practices of the previous experiences of developed countries, with the specific challenges of the global South. This creative combination is particularly important, as Big Data and associated technologies imply the emergence of a new power, due to the generation of new sources of knowledge building and decision-making capabilities. This modifies the balance of power that may be used to promote and even impose the interests of some groups over the interests of others. Just as they involve a new power, Big Data and associated technologies imply new responsibilities. It raises the question of the ethical ways of dealing with these new capabilities in a context of social inequality. In the face of this, the development of regulatory frameworks and the deployment of educational efforts are of the biggest importance: it is necessary to plan with the knowledge that these societies are not homogenous and contain large gaps in capabilities.

Limits and possibilities of the adoption of new information practices and technologies in dialogue with the preexistent context

The general challenges and risks created by the adoption of Big Data technology are amplified and specified by the context in which that adoption takes place. One of the main characteristics of developing countries is the lower prevalence of Big Data sets than of smaller, simpler datasets or situations with no data available. The distinction between Big Data and what we could call "small data" puts the latter one in a category born from contrast: what we now could consider "small data," would have been just "normal" data previously (Miller, 2010). Similarly, what today is considered "Big Data" could stop being considered that in the future when bigger and more complex datasets became available, updating what is needed to scale up its operation (Marz & Warren, 2015).

Compared to Big Data, small data is simple and can be handled by traditionally available tools like spreadsheet software or in-memory analytics. While handling Big Data requires, for example, the use of distributed file systems (HDFS in Hadoop's case), implies connectivity and coordination needs (Marz & Warren, 2015), small data can be processed using individual and standard resources, which do not require additional infrastructure (Kitchin & Lauriault, 2015). Based on that, it may seem that the optimum situation would be to upgrade from small data to Big Data and reap the benefits of the scale and power of analysis it allows. However, it is necessary to consider the context in which information is managed: while the transition is from small data to Big Data in developed countries, in developing countries the transition may be from having no available data to at least have some data at their disposal.

Data does not need to be big for it to be valuable. Vigilance surrounding different aspects of small data is still necessary; it is important to consider its purpose, its design, its collection, how privacy is handled, and what kind of accountability exists, etc. Big Data does not erase those questions, it amplifies them, and as it implies real-time or near real-time processing of a large flux of information, those questions must be continuously addressed. In the Peruvian context, it is important to clarify that greater access to relatively new data technology does not necessarily involve Big Data, especially considering that less than 60% of the population has access to the internet (Peruvian National Institute of Statistics and Informatics, 2019). For example, the use of satellite images for precision agriculture remains uncommon among agricultural producers even if the relevant data sets are not necessarily big – since updating frequency and level of detail are limited – despite requiring an enormous technological deployment. By comparison, simpler but more frequently updated data – like the one produced by monitoring sensors – may allow more tailored decision-making by providing information with greater detail about the situation. The scaling and connecting of technology through the Internet of Things is, for example, what can initiate the transition to Big Data.

However, although Big Data adoption is seen positively in the current global context, it has its share of potential problems. For instance, if one wants to decide about matters that impact the general well-being of a population, it may be better to have a small representative sample than to have access to a giant dataset, which may be affected by selection bias. It is not unusual to end up possessing large datasets built with questions in mind that differ from what is considered of interest later, precisely because new technologies sometimes allow for automatic, frictionless data collection, potentially creating an inadequate data set to answer those new questions. It may be better to have small data sets precisely and carefully built to answer the question of interest than to have a Big Data set that answers a considerably different question. Big datasets built around consumption decisions are conditioned by people's spending levels, potentially excluding individuals that do not spend beyond a certain threshold. Later, that same data set cannot be reliably considered a valid source to produce insights about the general public. It is necessary to have contextualized and adequate data handling in cases where that same data is later used to train models and algorithms that have consequences in different fields. Considering context is particularly important in cases of public-private collaboration that includes sharing information (Richterich, 2018), because the intention and logic that guided data generation and the selection of relevant variables in the private sector is not necessarily the same as or even compatible with the basic orientation of a public policy effort.

In the Peruvian case, sub-national differences in public service access and connectivity are also expressed by the fact that any implementation of new data generation platforms must consider coverage limitations and how this may create bias. For example, if automatic data integration mechanisms were implemented in areas like public health or education, the access gap between urban and rural areas would mean much more data is available about urban areas than rural areas, thus potentially over-representing the problems that urban areas face when establishing future public agendas. For example, according to official statistics, approximately 60% of the urban population has access to the internet, while it is lower than 25% among the rural population (Peruvian National Institute of Statistics and Informatics, 2019).

Data is a valuable resource because it generates multiple optimization opportunities in diverse areas. In the public health sector, it facilitates the monitoring and decision-making regarding both individual and collective cases: for example, decision-making regarding public policy based on epidemiological analysis, or changes in the medical insurance and medical care offering. In agricultural activity cases, data availability allows for better resource management (e.g. optimize how fertilizers, water, and pesticides are used), contributing to improved productivity and diminishing environmental impact. In the financial sector, the ever-increasing availability of bigger datasets is used to train more precise credit risk models, reducing the need for informal unsupervised credit agents with higher interest rates than the formal systems. In general, the rise of Big Data goes hand in hand with the parallel upsurge of new ways to analyze information and numerous new and more powerful techniques and algorithms. The application of complex neural networks and deep learning techniques has allowed the development of faster image recognition systems that make it possible to explore new frontiers and increase the scale of analysis in an unprecedented manner.

The confluence of Big Data and these multiple new ways of analyzing information, notably artificial intelligence, led to the emergence of intelligent systems (Stahl & Wright, 2018). They are capable of responding much more quickly to challenges from diverse sectors and are able to continuously refine the models that guide their actions, they attempt to automate as much of the process as possible with available resources, and they are able to predict future needs by analyzing observable patterns in the data registered by the system, complementing it with other data sources available to the platform holder. The use of new analytical tools also makes it possible to deploy smart system logic even around small data, such as in adaptive learning systems management, like those that use such techniques as Bayesian knowledge tracing (Hawkins, Heffernan, & Baker, 2014) for educational purposes. Automatic data generation allows for much more flexibility and detail when adapting to the needs of each student, while building data sets for aggregated analysis of collective student performance trends at the same time. It is necessary to discuss what the acceptable limits are to the application of these new technologies. Technology creates opportunities, and an external sphere must establish what is considered a legitimate use of those tools and what is not.

Some ethical challenges and dilemmas for public policies around Small and Big Data in heterogeneous Peru

In Peru, there are sustained gaps in access to new technologies, but also in access to more traditional societal elements such as quality education opportunities, appropriate health coverage, and others. These gaps are both the result and the cause of different paths available to Peruvian citizens. Policies that seek to promote a respectful, mutually beneficial relationship between the general public and the actors capable of producing, collecting, and using information – including the government – must take into account how ill-prepared large portions of the population may be to deal with the intricacies of data's life cycle and its implications. In order to make the most of the opportunities created by the emergence and use of new technologies like Big Data and Artificial Intelligence, those policies must consider the specific challenges faced by Peruvian society in the face of its internal inequality. The following is a summary of the three main challenges faced by the public administration in the adoption of Big Data and associated technologies in Peruvian society.

First challenge: ensuring that the adoption of new technologies increases opportunities for citizens while mitigating its impact on the expansion of social asymmetries

The diversity of conditions in Peruvian society is expressed in the multiplicity of special interests and perceived possible courses of action. The adoption of new technologies is not an exception to this disparity. Thus, the use of small data, Big Data, and associated technologies acquires different characteristics and generates bigger synergies for some actors than for others.

Inadequate education characterizes the Peruvian context. In the most recent OECD survey of adult skills (OECD, 2019), Peru ranked the lowest among all participating countries and economies, which included Chile, Ecuador, and Mexico. Only 6.6% of Peruvian participants between 16 and 65 years old showed high-level problem-solving skills in technology-rich environments, while 80.1% were low performers in literacy and/or numeracy. As for students, the most recent national standardized test, from 2018, showed that only 16% of second-year high school students reached their expected level in reading, and only 14% reached their expected level in mathematics. This trend is exacerbated in rural areas, with barely 3% of students achieving what was expected of them in reading and the same percentage in mathematics, and only four regions out of twenty-five showing above 20% of students with a satisfactory understanding of those crucial topics (Peruvian Ministry of Education, 2018).

In the face of the lingering existence of these more traditional challenges, it is necessary to understand that Big Data is not the only technology that can be used to promote the narrowing of social gaps. There remains the need to increase the skill levels used in more traditional elements, using technology like small data and offline tools (this is quite important, considering the limited access to the internet among Peruvian rural population, shown in the previous section). This could have a greater impact for major parts of the population, increasing their chances of also being able to extract value from data. For example, some authors point out that the value produced as a result of the more fluid connection of multiple small data sets can be even bigger than the value product of the mining of Big Data sets (Pollock, 2013).

As Taylor et al. (2014) points out, "Big data is not for everyone or every purpose. There are many circumstances where small data or non-digital approaches are most important for social change, particularly in lower-income environments where digital communications are still a relatively new phenomenon." It is necessary to have data-aware public managers and authorities, with enough knowledge to understand that it will be necessary to establish different strategies and use different tools for different groups and contexts. Preparing much of the Peruvian population to deal with future opportunities emerging from the use of new technologies requires a continuation of the efforts oriented to solve more basic, traditional problems (e.g., the lack of quality education), which comprise Peru's historical specificity.

Second challenge: finding an adequate balance between data generated about citizens and the protection of related privacy and security rights, in a context marked by a low public capacity to monitor the adequacy of data management practices and to enforce the law

In addition to the general challenges and risks associated with Big Data, as mentioned previously, the adoption of this type of technology in developing countries involves specific factors that complicate the usual challenges. For example, there is the question of the costs of not employing information, if such information threatens the right to privacy or some other civil right. This question is amplified by the sense of urgency associated with the need to find solutions to problems that cause widespread suffering. Thus, a dilemma: at what point does it become acceptable to deploy or to allow the deployment of these technologies even if they entail risks to some members of society, if it is perceived that the technology may help those or other members of society? What are the acceptable trade-offs? This is just one of quite a few questions that arise when speaking about how new opportunities occur among pressing problems, but one is enough to show that solutions are not simple, and that technology alone cannot solve ethical conundrums. It is different to discuss the limits that can be imposed on the use of technologies when that discussion occurs in a context of relative prosperity than when it occurs in a context with pressing needs and understandable social demands and dissatisfaction.

While the Peruvian Protection Law (Data Protection and Privacy Law, 2011) establishes opt-in logic that makes it necessary to obtain permission to register and use third-party information, in practice, people don't necessarily take the time to carefully analyze the implications of granting access to their data, and often hastily approve data collection and processing. Currently, the most frequent violation of this law according to members of the corresponding national authority is related to the lack of information provided to citizens by organizations about what kind of treatment their data will receive:

> The most frequently detected infraction is related to the issue of consent to the treatment of personal data. Organizations do not comply with the security measures in this issue and the most frequent finding in our inspections this year was that organizations do not comply with their duty to inform. The citizen has the right to know what will be done with his personal data, what the treatment will be, and how and who is going to do it; this information must be given to the citizen by any public or private organization that performs personal data treatment. What we have identified is that this information is not provided to the citizen (Huamán 2019).

Because of the presumably low level of data literacy among the population (since data literacy requires both numeracy and literacy skills), it is arguable that even if organizations would comply with the regulations regarding providing citizens with information about what kind of treatments would be applied to their data, in many cases people wouldn't be able to effectively assess the magnitude of what they accept by giving their consent. Besides, even if they would find acceptable the use of their anonymized data, it is highly unlikely that the organization that requests permission to use the data would explain that there are ways of reverting that anonymization later (de Montjoye et al., 2015; Huang et al., 2019; Rocher et al., 2019) and merge the data with some other datasets, for example, through probabilistic record linkage (United States General Accounting Office, 2001). Considering this, it would be advisable to continue developing a regulatory framework for the production and use of information, marked by the intent to protect the user, a framework that attempts to minimize unintentional exposure to undesirable data handling practices.

This is even more important when dealing with mandatory participation in some data collecting system. For example, the adoption of electronic health records is not something subject to individual choice, but a measure that the public healthcare system establishes and promotes. While the potential benefits for the patient are clear (getting better healthcare from the integration of all the relevant information in an easily searchable, secure environment), this is ultimately a promise that entails the patient trusting his information to the security mechanisms implemented in the system, and accepting these mechanisms as reliable. It is known that a much more experienced system, such as the UK's National Healthcare System has been subjected to ransomware attacks (Ghafur, Grass, Jennings, & Darzi, 2019) and has suffered data breaches due to something as trivial as a poorly composed email (Campbell, 2019) or due to inadequately coded data handling software ("NHS data breach affects 150,000 patients", 2018).

With all this in mind, and acknowledging the encouraging development of a data protection legal framework in Peru, it is necessary that authorities carefully ponder the most adequate timing and mechanisms for implementing new information technologies, especially considering the limited capacity of the public in monitoring data management practices, and the difficulties in enforcing the law (both due to scarce resources and to the high level of informality present in the Peruvian economy). Legal reform will not be enough by itself; it will be necessary that the strategic decisions adequately consider the context which makes it indispensable to carry out actions to educate the population on the importance of these issues.

Third challenge: promoting the availability and discoverability of public information even if it results in greater scrutiny and generates political costs which may inhibit risk-taking by decision-makers

The improvement of data-awareness and data-literacy among the Peruvian population is much more likely if its promotion emphasizes how information can be key to tackling issues that are considered a priority by the citizens. Two of the main issues are the fight against corruption and the related need for greater transparency and accountability of administrators. Thus, it is desirable to provide, for example, friendly access that makes it possible to track and corroborate the adequate provision of goods and services to the public.

In this context, the potential of data journalism stands out as a possible mechanism to mediate between citizen demands and data-sharing platforms. For this to happen, it would be necessary to provide journalists with tools to explore bigger and more complex data sets. This would not negate the agency of citizens, but it would complement it with the use of more advanced resources, like the potential use of application programming interfaces (APIs) that grant access to applications and databases through stable mechanisms established in documented protocols, making them programmatically usable, therefore favoring data analysis at massive scale.

Two data journalism portals, OjoPúblico and Convoca, have received international awards for showing the extent and characteristics of corruption in the country and its links with international networks. Inter American Press Association awarded Convoca the first place in data journalism category for its web application "Vía Sobrecosto" ["Cost overrun road"], that allowed to explore the cost overrun in public works projects due to corruption ("SIP exalta la labor del periodismo nicaragüense al anunciar los premios a la excelencia periodística 2018", 2018). OjoPúblico was awarded with the 2016 LatinAmerican investigative journalism prize by The Institute for Press and Society and Transparency International for its web application "Memoria Robada" ["Stolen memory"], that allowed to explore international networks of illicit trafficking of LatinAmerican cultural property ("OjoPúblico gana Premio Latinoamericano de Periodismo de Investigación", 2016); and also won the Sigma Awards 2020 in the inno-vation category for "Funes: an algorithm to fight corruption," that "using data scraped from five public databases... analyzed hundreds of thousands of Peruvian public procurement documents [and] using a linear model, it combines 20 risk indicators – such as recently founded contractors or uncontested bids – to flag potentially corrupt contracts" ("The 2020 Sigma Awards", 2020).

According to the OURdata index, built by OECD (2018) for measuring government efforts in open data implementation, Peru obtains half of the possible points for Openness, Usefulness, and Re-usability of government data. Despite this, an increase in the level of digitalization of Peruvian government information is imperative. This would allow for consolidat-ing incipient public information repositories, which currently have scarce and mostly outdated information. The task involves not only digitizing information and storing it, but also designing and implementing adequate mechanisms for using and sharing it, especially if we consider the low level of familiarity of a big part of the Peruvian population with this type of resource. The deployment of visualization portals is advisable, because it could increase the probability that the information ends up meaning some-thing for the citizen, making data more actionable and a better resource for decision-making. Some interesting cases are the website AtuServicio ["At your service"] in Uruguay (Uruguayan Ministry of Health, 2019) that allows the comparison of key indicators between different healthcare estab-lishments, and the Chilean website DataChile (Chilean Ministry General Secretariat of Government, 2019) that integrates and visually presents an important quantity of open government data, emphasizing integration and utility. Finally, it is important to continue the implementation of open data because the more visible and complete this data is, the greater the opportuni-ties to increase the visibility of disparate interests and needs, and to include them in the public discussion, promoting the participation of sub-national authorities and of the population that comprises this very diverse country.

Conclusion

The adoption of Big Data along with related analytical practices and tech-nologies in Peru's case – as in other developing countries – creates oppor-tunities to answer old and new problems, giving new tools to tackle the need to improve living standards. Making the most of those opportunities will require data-aware and diversity-aware officials and authorities who are capable of creating and deploying diversity-aware public policies that adapt these new tools to the highly unequal national context considering its historical specificity, and creatively tackle the mechanisms that underlie the current disparities, with the lack of opportunities (e.g. access to basic services and quality education opportunities) as one of the main limiting factors.

It will be necessary to deploy diversity-aware public policies for promoting a respectful and mutually beneficial relationship between the general public and actors (individual and collective, public and private) capable of producing, collecting, and using data. Given how ill-prepared big portions of the Peruvian population are to deal with the intricacies of data's life cycle and its implications, a safety-first approach to data collection and handling should be promoted in order to minimize the exposure of citizens to undesirable practices. In that context, small data's importance remains and evolves. Only by increasing the population's general data awareness and literacy will it be possible for citizens to be in a stronger position to demand the ethical production and use of information and its derived power, with an emphasis on protecting their rights to privacy and security while increasing government transparency and accountability. The derived ethical challenges imply defining what trade-offs are considered acceptable, what limits are fixed, what sectors are prioritized, and what principles are considered key guiding ideas. Far from being an exclusively technical discussion, the dialogue around the adoption of new information technologies like Big Data needs to consider different ethical visions. Utilitarianism, Kantianism, virtue ethics, and other frameworks (Herschel & Miori, 2017) are not things of the past, disconnected from the impressively fast technological evolution; they are raw materials that make it possible to take a stand and communicate it. For new information technologies to effectively contribute to improving living standards and reducing suffering in society, ethical reflection will be much needed to avoid inflicting the needs or preferences of the many on the rights of the few and vice-versa.

References

Campbell, D. (2019, September 6). NHS gender identity clinic discloses email contactsof 2,000 patients. *The Guardian*. Retrieved from https://www.theguardian.com/society/2019/sep/06/nhs-gender-identity-clinic-discloses-email-contacts-data-breach

Chilean Ministry General Secretariat of Government. (2019). DataChile. Retrieved October 31, 2019, from DataChile website: https://es.datachile.io

Data Protection and Privacy Law., Pub. L. No. 29733 (2011).

de Montjoye, Y.-A., Hidalgo, C. A., Verleysen, M., & Blondel, V. D. (2013). Unique in the Crowd: The privacy bounds of human mobility. *Scientific Reports, 3*, 1376.

de Montjoye, Y.-A., Radaelli, L., Singh, V. K., & Pentland, A. S. (2015). Unique in the shopping mall: On the reidentifiability of credit card metadata. *Science, 347*(6221), 536–539. https://doi.org/10.1126/science.1256297

Ghafur, S., Grass, E., Jennings, N. R., & Darzi, A. (2019). The challenges of cybersecurity in health care: The UK National Health Service as a case study. *The Lancet Digital Health, 1*(1), e10–e12. https://doi.org/10.1016/S2589-7500(19)30005-6

Hand, D. J. (2018). Aspects of Data Ethics in a Changing World: Where Are We Now? *Big Data, 6*(3), 176–190. https://doi.org/10.1089/big.2018.0083

Hawkins, W. J., Heffernan, N. T., & Baker, R. S. J. D. (2014). Learning Bayesian Knowledge Tracing Parameters with a Knowledge Heuristic and Empirical Probabilities. In S. Trausan-Matu, K. E. Boyer, M. Crosby, & K. Panourgia (Eds.), *Intelligent Tutoring Systems* (pp. 150–155). Cham: Springer International Publishing.

Herschel, R., & Miori, V. M. (2017). Ethics & Big Data. *Technology in Society*, 49, 31–36. https://doi.org/10.1016/j.techsoc.2017.03.003

Huamán, G. (2019, February 12). Autoridad de Transparencia y Datos Personales prioriza interés público y libertad de información. Retrieved October 31, 2019, from Ojo Público website: https://ojo-publico.com/975/autoridad-de-transparencia-y-datos-personales valora-interes-publico-y-libertad-de-informacion

Huang, Lin., & Lin. (2019). Data Re-Identification—A Case of Retrieving Masked Data from Electronic Toll Collection. *Symmetry*, *11*(4), 550.

Jurkiewicz, C. L. (2018). Big Data, Big Concerns: Ethics in the Digital Age. *Public Integrity, 20*(sup1), S46–S59. https://doi.org/10.1080/10999922.2018.1448218

Kitchin, R., & Lauriault, T. P. (2015). Small data in the era of big data. *GeoJournal*, 80(4),463–475. https://doi.org/10.1007/s10708-014-9601-7

Laney, D. (2001). *3D Data Management: Controlling Data Volume, Velocity and Variety* (Gartner Report, p. 3). https://blogs.gartner.com/doug-laney/files/2012/01/ad949-3D Data-Management-Controlling-Data-Volume-Velocity-and-Variety.pdf

Lipworth, W., Mason, P. H., Kerridge, I., & Ioannidis, J. P. A. (2017). Ethics and Epistemology in Big Data Research. *Journal of Bioethical Inquiry, 14*(4), 489–500. https://doi.org/10.1007/s11673-017-9771-3

Marz, N., & Warren, J. (2015). *Big data: Principles and best practices of scalable real time data systems.* Manning

Miller, H. J. (2010). The data avalanche is here. Shouldn't we be digging? *Journal of Regional Science*, 50(1), 181–201. https://doi.org/10.1111/j.1467-9787.2009.00641.x

Mittelstadt, B. D., & Floridi, L. (2016). The Ethics of Big Data: Current and Foreseeable Issues in Biomedical Contexts. *Science and Engineering Ethics*, 22(2), 303–341. https://doi.org/10.1007/s11948-015-9652-2

NHS data breach affects 150,000 patients. (2018, July 2). BBC News. Retrieved from https://www.bbc.com/news/technology-44682369

OECD. (2018). *Open Government Data Report: Enhancing Policy Maturity for Sustainable Impact.* https://doi.org/10.1787/9789264305847-en

OECD. (2019). Skills Matter: Additional Results from the Survey of Adult Skills. OECD. https://doi.org/10.1787/1f029d8f-en

OjoPúblico gana Premio Latinoamericano de Periodismo de Investigación (2016). OjoPúblico. Retrieved from https://ojo-publico.com/339/ojopublico-gana-premio-latinoamericano-de-periodismo-de-investigacion

Peruvian Ministry of Education. (2018). *¿Qué aprendizajes logran nuestros estudiantes?* Retrieved from Ministry of Education website: http://umc.minedu.gob.pe/wp-content/uploads/2018/10/Informe-Nacional-ECE-2018.pdf

Peruvian National Institute of Statistics and Informatics. (2019). Estadísticas de las tecnologías de información y comunicación en los hogares – Trimestre: Julio – Agosto – Septiembre 2019. (No. 4). https://www.inei.gob.pe/media/MenuRecursivo/boletines/ticdiciembre.pdf

Pollock, R. (2013, April 25). Forget big data, small data is the real revolution. *The Guardian*. Retrieved from https://www.theguardian.com/news/datablog/2013/apr/25/forget-big-data-small-datarevolution.

Richterich, A. (2018). *The big data agenda: Data ethics and critical data studies.* London: University of Westminster Press.

Rocher, L., Hendrickx, J. M., & de Montjoye, Y.-A. (2019). Estimating the success of re identifications in incomplete datasets using generative models. *Nature Communications, 10*(1), 3069. https://doi.org/10.1038/s41467-019-10933-3

SIP exalta la labor del periodismo nicaragüense al anunciar los premios a la excelencia periodística 2018. (2018, September 5). The Institute for Press and Society. Retrieved from https://www.sipiapa.org/notas/1212684-sip-exalta-la-labor-del-periodismo-nicaragense-al-anunciar-los-premios-la-excelencia-periodistica-2018

Stahl, B. C., & Wright, D. (2018). Ethics and Privacy in AI and Big Data: Implementing Responsible Research and Innovation. *IEEE Security & Privacy, 16*(3), 26–33. https://doi.org/10.1109/MSP.2018.2701164

Steinberg, E. (2019). Big Data and Personalized Pricing. *Business Ethics Quarterly,* 1–21. https://doi.org/10.1017/beq.2019.19

Taylor, L. (2016). *Group privacy: New challenges of data technologies.* New York, NY: Springer Berlin Heidelberg.

Taylor, L., Cowls, J., Schroeder, R., & Meyer, E. T. (2014). Big Data and Positive Change in the Developing World: Big Data and Positive Change in the Developing World. *Policy & Internet, 6*(4), 418–444. https://doi.org/10.1002/1944-2866.POI378

Terry, N. (2014). Health Privacy Is Difficult but Not Impossible in a Post-HIPAA Data Driven World. *Chest, 146*(3), 835–840. https://doi.org/10.1378/chest.13-2909

The 2020 Sigma Awards (2020). Sigma Awards. Retrieved from https://datajournalism.com/awards

Uruguayan Ministry of Health. (2019). AtuServicio.uy. Retrieved October 31, 2019, from http://atuservicio.msp.gub.uy/

United States General Accounting Office. (2001). *Record linkage and privacy: Issues in creating new federal research and statistical information.* https://www.gao.gov/new.items/d01126sp.pd

Vines, P., Roesner, F., & Kohno, T. (2017). Exploring ADINT: Using Ad Targeting for Surveillance on a Budget - or - How Alice Can Buy Ads to Track Bob. *Proceedings of the 2017 on Workshop on Privacy in the Electronic Society - WPES '17*, 153–164.https://doi.org/10.1145/3139550.3139567

Zwiegelaar, J., & Stylos, N. (2018). Is Big Data the next Big game changer? Impact on Customer services, Marketing and Ethics. *BAM 2018 Conference Proceedings.*

4   Open government, dilemmas, and innovation at the local level

Comparing the cases of Austin, Buenos Aires, and Madrid

Edgar A. Ruvalcaba-Gomez, Soledad Gattoni, and Raymond W. Weyandt

Introduction

The activities and roles of governments have changed considerably due to a new reality in which information flows in large quantities and in real time. In this sense, open government is a new model that seeks to respond to a new reality and give direction to the public sector that demands greater cooperation between government and society. The structural modernization that is linked to open government is due to a change of era, to the emergence of a "network society," and to innovation in new government practices linked to the potential use of new technologies.

In public administrations that have assumed the open government model to develop different public policies, it is possible to observe a variety of approaches to its implementation. In this sense, this research is based on the hypothesis that there are many political discourses that revolve around open government and its application as a new way of governing. However, these discourses and ideas they assume are very different. Therefore, it is important to explore the experiences from case studies that allow us to know the different perspectives assumed and those conceptual gaps that constitute the materiality of open government, as well as reflect on the dilemmas facing the implementation of this new public management model.

This research analyzes the institutional development and the results of open government in three subnational governments: Austin (United States), Buenos Aires (Argentina), and Madrid (Spain). Three local administrations with different characteristics in the political, social, and economic context. However, the three cities are part of the Open Government Partnership (OGP) Local[1] Program. OGP is a multilateral initiative that aims to secure concrete commitments from national and subnational governments to promote open government, empower citizens, fight corruption, and harness new technologies to strengthen governance. In this sense, the three cases have a frame of reference established by OGP that homogenizes criteria which configure the comparable analytical conditions. This work establishes a research question that guides the study: What visions and perspectives have framed the open government approach in the local governments of Madrid,

Buenos Aires, and Austin? In order to answer this research question, we use the Open Government Perspective Model (OGPM) proposed by Ruvalcaba-Gomez (2019), which is useful to analyze the perspectives through which open government develops in public administrations.

The research recognizes the need to deepen studies at the local level, capable of accounting for the adoption of open government public policies in governments of first contact with citizens. In this situation, local public administrations constitute an area of opportunity to inquire into the issue, not only because of the lack of studies at this level of government, but because they represent the administrative unit mostly linked to citizen demands of daily nature (Blanco and Gomà 2003; Conradie and Choenni, 2014; Criado and Ruvalcaba-Gómez, 2018; Grimmelikhuijsen and Feeney, 2017; Sivarajah et al., 2015). Indeed, local governments have developed some of the most relevant experiences in open government at the international level, as evidenced by the recent incorporation of twenty governments to the sub-national program of action plans of the OGP, as well as the increasing amount of studies on this level of government in international literature.

The development of this research is organized in different sections. The following section presents a Literature Review and the Analytical Framework in which we define some basic notions about open government, and we also establish the analytical strategy used for research. In the third section we develop our research methodology, where we refer to research methods and techniques, as well as other features related to the instrumentation of the analysis. The fourth section presents the research findings, divided into the three case studies and a comparative analysis. Finally, we present some final conclusions and implications of our research.

Literature review and analytical framework

Open government is a concept utilized by different actors and social sectors. Many politicians, public managers, civil society leaders, and academics have referenced the concept of open government in speeches and publications, mainly highlighting its potential and its implications. However, there is a problem with the perception of open government and how it is being implemented. Due to the recent reemergence of the concept and its polysemic characteristic, open government is perceived differently by different actors, which is reflected in the government actions of public administrations that assume policies of institutional openness.

Within the topic of open government, there is debate over the structure of the relationship between the government sector and society. Today, we find a plurality of ideas and definitions about open government from both academic authors (Abu-Shanab, 2015; Ganapati and Reddick, 2012; Gascó, 2014; Lathrop and Ruma, 2010; Meijer et al., 2012; Noveck, 2015; Wijnhoven et al., 2015) and international organizations (IDB, 2016; ECLAC, 2016; European Commission, 2016; OECD, 2016; OAS, 2016; OGP, 2011, Red

Gealc, 2016).There is a certain level of ambiguity when we talk about open government (Yu and Robinson, 2012). The wide variety of conceptual approaches clearly shows that open government is a polysemic term which requires an in-depth academic analysis.

The basic elements, or pillars, of open government represent a permanent debate about the perspectives of researchers and the studies that support them. Some authors refer to open government as a topic under development and with little research (Lee and Kwak, 2012; Criado, Ruvalcaba-Gómez, 2016; Ruvalcaba-Gomez and Valenzuela, 2018). However, several studies have led to the inclusion of a more complex conceptualization where many of them converge in considering three pillars: transparency, participation, and collaboration (Lathrop and Ruma, 2010; Lee and Kwak, 2012). Other studies carried out in recent years indicate that not all of these maintain similar directions regarding the elements of open government (Abu-Shanab, 2015; Williamson, and Eisen, 2016; Wirtz et al., 2015).

To establish an analytical framework that is useful for us to know the approaches to open government perspectives, we use the "Open Government Perspectives Model" (OGPM), developed by Ruvalcaba-Gomez (2019). The OGPM serves as a frame of reference to categorize the perspectives of open government that are useful to understand the vision of governments in relation to the opening of institutions. This model has antecedents in the work of Ruvalcaba-Gómez, Criado and Gil-García (2017) which already presented the emergence of the three perspectives and applied it to a specific case. The reference literature to the analytical model shows us that the three analytical dimensions emerge from a factor analysis linked to the theoretical analysis. These dimensions symbolize the perspectives with which policies are normally assumed to be open government initiatives:

- Democratic values of co-responsibility
- Technological innovation
- Availability and access to information

Thus, the OGPM archetype is configured. This model is useful for understanding and explaining how open government policies are constituted in a public administration that is developing, adopting, and implementing actions and programs designed to increase transparency, civic participation, public accountability, and other forms of "openness." The OGPM is constructed from thirteen concepts that arise from the review of the literature and that are analyzed empirically in a study by Ruvalcaba-Gomez, Criado and Gil-Garcia (2017). The model is based on thirteen concepts that make up the three dimensions mentioned: a) Democratic values of co-responsibility: *democracy, co-creation, collaboration, citizen participation,* b) Technological innovation: *electronic government, interoperability, new technologies, social networks, smart cities,* and c) Availability and access to information: *access to information, accountability, open data, transparency.*

The perspective or dimensions used in the analytical model are supported by multiple theoretical and empirical studies that have been addressed and discussed in the academic literature regarding open government. Several authors have built a large debate about what is and what is not open government, its elements, concepts, dimensions, and categories, so that it is now possible to solidly support the three dimensions of the OGPM. The following explains what each of the perspectives mentioned according to Ruvalcaba-Gomez (2019).

Perspective of democratic values of co-responsibility

At present, democracies have adopted a series of ideas and values that constitute a representative system that assumes the will of the majority of a society. In this logic, the responsibility for public affairs has been shared by society and the government, giving rise to the configuration of a perspective of democratic values of co-responsibility. Co-responsibility, as a value, is assumed from an open government model and configured from a series of concepts such as democracy, collaboration, citizen participation, and co-creation.

On the one hand, authors such as Janssen, Charalabidis, and Zuiderwijk (2012) have categorized the benefits of open government into a "political and social" dimension in which they group various elements such as transparency, democratic accountability, empowerment, and citizen participation or the process of creating policies. These elements are directly linked to the "perspective of democratic values of co-responsibility," understood as the collective construction of public decisions and shared responsibility between the governmental sphere and civil society to make decisions, co-create strategies, and implement public policies.

To reinforce this perspective, Wijnhoven, Ehrenhard, and Kuhn (2015) established three objectives of open government practices that are closely linked to the democratic vision of co-responsibility. This link is due to a collaborative vision of citizens as a source of policy proposals and the idea of citizen innovation as substantial elements of open government. Another important study that is related to the perspectives supported by the GMO is the work of Williamson and Eisen (2016), in which they classify three governance processes from an open government approach. One of these processes involves interventions aimed at broadening public engagement and participation. This is consistent with the concepts of participation, collaboration, and democracy linked to the perspective of democratic values of co-responsibility.

Perspective of technology innovation

Technological innovation is a key factor for development within public management, which involves generating products and services from new citizen-oriented elements. In this sense, it is possible to associate the perspective of technological innovation with the open government model when it is linked to concepts such as electronic government, new technologies, interoperability, smart cities, and social networks. Within the study of open government, the use of technologies is a common denominator. Gascó (2014) highlights the relevance of the use of technologies as an essential element in the construction of open government. For its part, Criado (2016) points out that public innovation is the path that every open government administration must follow if it aspires to be intelligent and respond to today's challenges of governance.

Noveck is one author who has become a reference in the perspective of "technological innovation." Her work highlights the potential of the intelligent use of technologies for open government. Noveck developed several studies on smart cities and their relationship with the open government that constitute a fundamental basis for further analysis (Noveck, 2015). The author talks about ways to potentiate and incorporate knowledge from citizens to government action. This knowledge has great value to solve public problems. Another reference that supports the technological innovation factor is the open government declaration made by the OGP, which establishes that increasing access to new technologies is an essential element for government openness and the construction of an open government model (OGP, 2011).

Perspective of availability and access to information

In recent years, governments have expressed interest in strengthening institutions and regulatory frameworks in order to guarantee access to public information and promote government transparency in response to citizen demands for increased accountability. In this sense, this establishes the open government perspective of availability and access to information, which comprises concepts such as open data, access to information, transparency, and accountability. Sandoval-Almazan (2011) proposes a model of two perspectives of open government. They identify the "back door" perspective, oriented towards the old idea that the government owns public information, and contrast it with the idea of an "entrance door," which represents the opening that leads to an open government.

The line of research assumed by Abu-Shanab (2015) connects the elements of accessibility to information and accountability to Open Government, particularly in relation to the perspective of availability and access to information. Similarly, the elements of transparency and access to information are backed by studies that link the regulation of transparency and rights of access to public information with open government (Dawes and Helbig, 2010; Jaeger and Bertot, 2010). On the other hand, there is a proposal to classify the governance processes in open government presented by Williamson, and Eisen (2016) understood as "initiatives to increase transparency."

Methodology

To obtain empirical evidence, we selected three case studies based on geographic representation criteria and participation in the OGP's Local Program: Austin (United States), Buenos Aires (Argentina), and Madrid (Spain). The cases cover three reference latitudes in the development of open government, with different political, social, and economic characteristics: Europe, North America, and Latin America. However, the cases are analyzed in the light of their participation in the OGP Local Program, which establishes a framework to promote policies of institutional openness among subnational governments.

Based on the OGPM, we establish three analytical dimensions: democratic values of co-responsibility, technological innovation, and availability and access to information. The three dimensions are associated with concepts and themes that are references in open government literature and that are useful for analyzing their presence and relevance in the cases analyzed (see figure 4.1). The concepts for analysis include access to information, smart cities, co-creation, collaboration, open data, democracy, electronic government, interoperability, new technologies, citizen participation, social networks, accountability, and transparency.

Each case was analyzed by a different academic researcher, who was appointed by the OGP's Independent Reporting Mechanism (IRM) to evaluate the development and implementation of action plans presented by each local government. In this sense, the investigators are open government experts and know each case thoroughly. The sources of information for the analysis presented in this research include the evaluation reports written by these researchers and the material used therein, such as interviews and

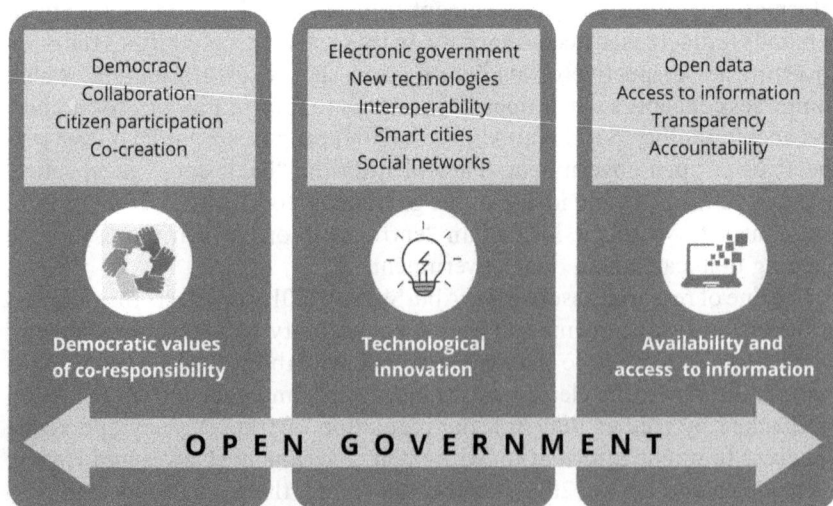

Figure 4.1 Open government perspectives model.

documentary analysis. Based on these sources of information, the researchers make a qualitative link between the discourses and documentary references to determine the weighting of the visions of governments based on the three analytical dimensions of the OGPM.

To develop the research process, the thirteen concepts mentioned serve as operational categories to classify the policies, political speeches, and their level of importance that they have to promote open government actions. In this way it is possible to determine the magnitude of each dimension or perspective of open government in accordance with the OGPM. It is worth mentioning that the information collection was carried out according to the schedule of the OGP Local Program. This schedule covered the first action plans at the local level, so the information was primarily collected between the months of May and December 2017.

Findings

This section presents the results of the application of the OGPM that constitutes an analytical instrument regarding the perspectives of open government. The objective is to present an analysis on the visions and perspectives that have framed the open government approach in the local governments of Madrid, Buenos Aires, and Austin. The OGPM serves as an analytical tool to understand the extent to which each local government assumes the three perspectives when developing open government initiatives.

Madrid case

The city of Madrid is the most populous in Spain and its capital located in the center of the Iberian Peninsula, in southern Europe. The municipality of Madrid is governed by the city council, which also consists of a representative corporation, chaired by the mayor, who is the highest executive authority. The Mayor of Madrid, Manuela Carmena, has assumed a substantive political commitment around the OGP, especially since the request to incorporate the city of Madrid into the pilot program of sub-national entities, which took place in the first half of 2016. Within the theme of open government, Madrid has included relevant issues related to transparency and citizen participation. Within the government team of the Madrid City Council, a leadership axis was established in the process of implementing the action plan before the OGP around the Government Area of Citizen Participation, Transparency, and open government. In order to analyze in greater depth, the configuration of the open government perspectives in Madrid, the dimensions of the OGPM are analyzed below.

Democratic values of co-responsibility

The perspective of democratic values of co-responsibility is present prominently in the Madrid case. This is due to the characteristics and principles that are more internalized by the determining actors in the construction of open government policies. These characteristics and principles are recognized in each of the conceptual elements and that are reflected in the actions of governmental openness. To configure the perspective of democratic values of co-responsibility according to the OGPM, the concepts of democracy, collaboration, citizen participation, and co-creation are included. These concepts are very present and in different media within the interviews carried out. In the case of Madrid, the concept of "citizen participation" stands out. This concept tops the list of concepts by becoming the most recurrent among the interviewees' discourse related to open government policy in Madrid.

The second most prominent concept is that of "democracy," this element is very recurrent among senior officials who are leading open government policies. An example of the way in which the interviewees expressed themselves in this regard can be found in the following statement "the open government policy in Madrid responds to a social demand that has become widespread throughout the last years and that has to do with a logical question of democracy." In this sense, a vision is assumed to include the citizen in municipal public affairs in order to legitimize his government and respond to a demand for real democracy. On the other hand, the concepts of "collaboration" and "co-creation" play an important role in referring to the interaction with civil society organizations and/or with international organizations that involved spaces for dialogue and joint work to include different ideas in the Government Action. The concepts are also important elements around citizen participation policies.

Technological innovation

The perspective of technological innovation has played a fundamental role within the open government policy in Madrid, in the sense that the concepts of electronic government, new technologies, interoperability, smart cities, and social networks have a presence in political and technical documents, and speeches. Within this perspective, the concept of "new technologies" stands out, the concept's relationship with the open government has its main connection when it is understood as a means to instrumentalize a series of public policy actions. In this sense, it is said that Madrid has the idea of innovating technologically or taking advantage of new technologies to innovate in democratic processes associated with decision making. It is possible to show that in Madrid there is a start-up of digital technology platforms that seek to reach a large part of citizens and overcome the digital divide.

In relation to "electronic government," a preferential link to the use of the Internet to offer services to citizens is shown. In this sense, Madrid makes use of free software for citizen participation policy, particularly to promote the "Decide Madrid" platform. As for the issues of "social networks," "smart cities," and "interoperability," there is a smaller presence. However, they are aware that it is a pending issue to potentiate the dissemination of policies. In this sense, the sources analyzed show an effort in dissemination and social networks.

Availability and access to information

Within the perspective of availability and access to information, it was possible to identify the relevance of certain elements related to the concepts of open data, access to information, transparency, and accountability. In these four elements, the presence of "transparency" stands out. Transparency is undoubtedly a concept highly associated with the open government in Madrid. In this area the greatest achievement has been the Transparency Ordinance of Madrid, which establishes a regulatory framework that guarantees higher standards of openness and some minimums in terms of access to information, open data, and accountability.

On the other hand, the element of "open data" is assumed as a policy related to transparency but that represents a more operational and useful resource for society based on its ability to generate new services. It is worth mentioning the relevance granted by the City Council of Madrid to open data, since these constitute one of the three aspects in which its open government policy is divided along with citizen participation and transparency. In relation to "access to information" is presented as a concept closely linked to transparency; However, access to information is a precondition for decision making. Finally, regarding the element of "accountability" in Madrid it is associated with the supervision and surveillance by civil society actors towards the government, this denotes that accountability mechanisms are perceived as an obligation for part of the municipal public administration.

Buenos Aires case

The City of Buenos Aires, officially called the Autonomous City of Buenos Aires, is the capital of the Argentine Republic and the main urban location in the country. The highest authority of the City of Buenos Aires is the Head of Government, who is head of the Executive Power of the City and has the role of administering the budget sanctioned by the legislature of the City of Buenos Aires (Legislative Power) and endorsing the decrees. The City of Buenos Aires also has a Judicial Branch with jurisdiction in criminal matters, neighborhood, offenses, contraventions, contentious-administrative, and taxation. The Head of Government of the City of Buenos Aires, Horacio Rodríguez Larreta, assumed strong political leadership within the framework of the OGP, since the City of Buenos Aires was incorporated into the pilot program of subnational entities in mid-April 2016.

The City of Buenos Aires' open government initiatives extend beyond OGP commitments and are framed in an open government ecosystem consisting of transparency, innovation, citizen participation, collaboration, and accountability initiatives. Particularly within the theme of open government, the City of Buenos Aires has included relevant commitments addressing the following issues: a) the aspiration to advance transparency and access to public information in the three branches of the State (executive, legislative, and judicial); b) the opening of public data on transport; education; and sexual and reproductive health; and c) the improvement of public services through the implementation of reporting mechanisms for infrastructure works in education, as well as in the provision of sexual and reproductive health services.

Democratic values of co-responsibility

"Co-creation" and "collaboration" are the concepts which stand out in the case of the City of Buenos Aires. Civil society organizations were able to observe, inform and influence the decision-making process on most of the commitments. According to the responses and perceptions collected during the IRM researcher's interviews, the relationship between the different stakeholders was productive, and of mutual collaboration and understanding. Although the general secretariat led the process of co-creation, collaboration between civil society and government representatives happened at all stages of the process: since the elaboration of the action plan, during the implementation, and monitoring of commitments. Minutes of the roundtable meetings as well as a first draft of the commitments were discussed and commented on virtually through an online repository. Moreover, when civil society representatives expressed difficulties when it came to attend face-to-face meetings, the multi-stakeholder forum agreed to maximize online collaborations and meetings at civil society organization headquarters to mitigate this situation and positively contribute to the co-creation process.

The second most prominent concept is "citizen participation." Four of the five commitments of the City of Buenos Aires action plan were relevant to the value of citizen participation, particularly including civil society organizations in the design of databases in the areas of public infrastructure works and access to sexual and reproductive health. Moreover, the action plan itself is framed within a broader "open government ecosystem"[2] that goes far beyond OGP action plans. Within this ecosystem, most of the initiatives include citizen participation in the design of public policies. One of these examples is "Buenos Aires Elige," a platform inspired by the "Decide Madrid" platform.

Technological innovation

Although all the commitments of the City of Buenos Aires action plan included the OGP values of technology and innovation in all its commitments. This perspective only plays a role as a contributor to foster access to information or citizen participation. The administration refers to "top-down innovation" showing how innovation is put at the service of the citizens themselves. Interoperability and social networks are the most important concepts of this dimension. Interoperability is connected to the establishment of open data standards and the publication of new databases. Social networks are used as a way to communicate with citizens and promote the work the City of Buenos Aires is doing in the field of open government and citizen participation. The concepts of "smart cities" and "electronic government" are not so frequently connected to the open government framework. This is because these public policy areas are driven by different agencies besides the general secretariat.

Availability and access to information

The most important concept within this dimension is "open data." During the implementation of its first action plan, the City of Buenos Aires released more than fifteen new databases in judicial, legislative, public works, transport, and health matters. At the same time, the establishment of a georeferenced platform on public works of education and the generation of reporting mechanisms in the area of sexual and reproductive health aimed to improve the quality of public services provided by hospitals and health centers in the city; as well as monitoring the progress of infrastructure works in public schools. All five commitments aimed to release information either with active or passive transparency measures. However, the concept of transparency is restricted to the release of access to information, in detrimental of other broader concepts of transparency associated with the implementation of anti-corruption mechanisms. Finally, the concept of "accountability" is the least present. Like in Madrid, the concept is associated with the surveillance by civil society actors towards the government. However, there is a lack of commitments relevant to this dimension. Closing the feedback loop with citizens, enforcing government actions, and incorporating mechanisms for citizens to denounce irregularities and strengthening public integrity are still pending.

Austin case

Austin is the administrative capital of Texas, the second most populous state in the United States. Austin, the eleventh most populous city in the United States, is home to nearly one million residents. The city government is administered by a ten-member city council, a mayor and a city manager. Each city council member is elected by residents of the district which they represent, while the mayor is elected at large, and the city manager is appointed by the city council and mayor. Austin was one of the first subnational governments included in the OGP's pilot program for local government. City leaders boast a proactive commitment to open government, often citing the city's prolific publication of data on public websites, participatory budgeting processes, and inclusion of citizen groups in decision making processes as evidence of the government's commitment to the principles of openness. Austin's open government efforts led to improvements in access to information and civic participation, along with the development of new technologies for citizens (Weyandt 2018).

Democratic values of co-responsibility

The depth of co-responsibility varies greatly across Austin's open government initiatives, especially those within the OGP portfolio. With no official central coordinator, there is no standard measure to ensure the co-creation of policies with community members. This did not prevent departmental initiatives from engaging with large segments of the community. Indeed, one deeply transformative initiative was the development of an equity assessment tool with which officials can analyze projects and budgets. The equity tool initiative is the clearest example of the city government's direct engagement of organized community members in the decision-making process. The Equity Office itself is a product of organized civil society.

Similarly, the Austin government's efforts to address the increasingly public problem of homelessness in the city intentionally incorporated a series of collaborative, citizen participation measures. City staff convened a council of individuals who were actively experiencing homelessness, representatives from local service providers, and city staff from departments addressing issues related to homelessness. In many ways, this research initiative was the first of its kind in Austin, a truly collaborative, inclusive. and direct conversation between a vulnerable community and the government entities committed to their service.

Amid a professed culture of openness, there are still plenty of government efforts in Austin that bypass a robust interaction with the community. Open government advocates in civil society and individuals within the city government continue to cite examples of the government's failure to apply the principles of openness, including closed door hiring processes for appointed officials and inconsistent adherence to transparency standards by the city's police department. This dichotomous adherence to open government principles draws the ire of transparency advocates as well as many government employees, as controversies rooted in secrecy create negative spillover effects for uninvolved staff and departments.

Technological innovation

Austin is known increasingly for its technology sector, so technological innovation is a natural component of any government program, especially new technologies. The city boasts a collection of interactive, prolific data portals. One of Austin's highest scoring OGP commitments was the development of a new online tool that allows residents to track progress and funding of projects in their neighborhood. Other online portals allow individuals to access data on capital projects and finance data. The city's official Open Data Portal houses more than 300 datasets from a variety of departments.

The city's prolific publication of data may be seen as an indicator of commitment to the principles of open government. A key opportunity exists in the area of interoperability, as departmental government data in Austin is rarely uniform across departments. Interview respondents during Austin's OGP assessment process repeatedly cited data uniformity and systems interoperability as key challenges to actionable data sharing between city departments. City officials have made a widespread use of social networks to publicize community forums and other events that encourage citizen engagement. In this sense, Austin actively attempts to increase civic participation through technology.

Availability and access to information

Austin's growing culture of open government is reflected in many successful efforts by city staff to enhance access to information, transparency, and accountability. The sheer quantity of data provided through the city's many public data portals is sufficient evidence of proactive open data policies. The city also adheres to Texas state public information laws and publishes data requested in accordance with the state Open Records Act. The city receives and publishes open records requests through a user-friendly online portal. Austin's open data portal houses hundreds of data sets on everything from city pools to dog parks to construction permits.

While publishing data is an important component of transparency, local transparency advocates continually cite inconsistencies in open government practices in some city departments, most notably the Austin Police Department (APD). APD receives the largest share of the city's annual operating budget, so a high level of transparency and accountability is expected by community leaders. City staff, including police department employees, admit a certain level of secrecy and disdain for public information requests within the police department. Citing further opportunities for improvement in transparency, city staff and open government advocates highlighted a controversial case of secrecy involving the hiring of the current city manager. These high-profile examples illustrate the ongoing opportunities for Austin's leaders to reaffirm their commitment to the principles of transparency and accountability across all city departments.

Comparative analysis and dilemmas

After analyzing the three cases based on the OGPM, we can contrast the most relevant characteristics of each case and determine the differences and similarities in relation to the three perspectives of the model. In this sense, we can confirm that the three cases assume the values promoted by the OGP: transparency, civic participation, accountability, and cross-cutting technology innovation and potential use of new technologies. On the other hand, the three cases show a lack of enforcement and consistency. In other words, there is no formal mechanism that guarantees the execution and application of open government strategies and government departments behave very differently from one another with regard to open government strategies.

Open government currently faces different dilemmas and issues that reflect a mismatch between what the theory indicates and what happens in practice. Different cultural, political, technological, and regulatory barriers create obstacles to implementing a model of openness in governments that address multiple practical aspects. A frequently recurring premise in the discourse on the implementation of open government is the cultural barrier that exists. Transparency, participation, and access to information policies, require many internal state cultural changes. Even the preceding analysis suggests a perception of disappointment in the ability of the open government movement to increase transparency and accountability.

Open government rhetoric has generated high expectations. These expectations exceed the outcomes realized by these new tools and through these new policies. In terms of citizen participation, there is still a great amount of progress to be made. Experts criticize the efficiency of decision-making processes under these slow-moving dynamics. This criticism posits that public management needs to make quick and accurate decisions and that consultation with the citizen, as is taking in the participatory processes in Madrid, naturally delays this process. This is a dilemma surrounding the extent to which participation takes and promotes direct or deliberative democracy.

Another dilemma has to do with the different understandings of democracy itself and the current changes in legitimacy and representation models. The case of Buenos Aires, for example, shows how open government rhetoric can be a means to its own ends as a platform for representatives to generate proximity and built trust with its citizens. Although the obvious answer would point to a balance, the dilemma has become a criticism by those who claim that execution is paramount, and listening is secondary. Following the analysis of citizen participation, there is also a criticism regarding the quality of participation by digital media, mainly in relation to the level of different types of involvement during the process. In certain cases, the implementation of participation mechanisms has caused conflicts instead of generating consensus and solutions. This is due to the nature of public consultations where different alternatives to solve or manage the same problem are submitted to public scrutiny.

An interesting criticism that deserves a deep analysis is related to the enforcement of open government commitments. Low levels of active citizen participation in democratic consultations is a constant challenge for open government reformers. This represents a paradox, as society demands simultaneously better democracy and participation while participating less and less frequently in the mechanisms of democracy. The positivist speeches of the same open government reformers have generated high expectations. The interviews in the Madrid case allow us to point out that some actors think that the idea that all transparency and participation efforts do not change anything in depth is certainly present in some citizen viewpoints.

A final dilemma concerns the lack of accessibility for marginalized communities to the benefits of open government reforms. New technologies and data portals mean little to citizens if they do not possess the resources to access them. Similarly, if data is produced in a single language, when local communities speak a variety of languages, then the data is for limited, exclusive use. This is contradictory to the principles of open government. Governments must, therefore, implement mechanisms within open government programs to ensure equitable outcomes for all citizens, especially those in systemically marginalized communities.

Conclusion

After analyzing the results derived from the research process, it is possible to highlight and reflect on some findings. We recognize the presence of an open government rhetoric that revolves around the issues of transparency, citizen participation, accountability, and co-creation. These results affirm the importance of incorporating the three perspectives of the OGPM into the implementation of open government initiatives to achieve maximum benefit. They also lend new insights to the ongoing discussion of the political, technological, legal, and social barriers involved in adopting open government commitments. Finally, this analysis initiates a discussion of the ethical dilemmas raised by the adoption of an open government framework, including the use of new technologies, open data, citizen participation, and access to information.

In the case of Madrid, the most outstanding actions of the municipal government in relation to the issues of transparency and citizen participation are fundamentally the "Transparency Ordinance" and the "Decide Madrid" platform. It is possible to conclude that the perception that transparency is diminished in the face of the role of citizen participation actions. It is notable that a communicative effort has been made in the government of Madrid in the part of citizen participation within the open government policy. This effort is not reflected in the same way in relation to the new transparency policies. In the city of Madrid, progress is recognized in terms of transparency by the city council. However, open government policy is focusing dissemination efforts on citizen participation. Although the city council promotes open government based on three pillars (transparency, citizen participation, and open data), in reality the policy only addresses them in the first two issues. Open data works as an operational action within the issue of transparency.

In the case of Austin, a community-designed equity assessment tool, a council on homelessness that includes directly impacted residents and a collection of open data portals represent strong cases of the city government's commitment to principled open government. Within each city department, it is common to find staff and initiatives that proactively embrace transparency and accountability. A strong standard of openness exists across most city departments. Inconsistencies persist, however, and high-profile controversies involving closed door proceedings, cover-ups, and secrecy threaten Austin's reputation as an open government pioneer. Austin's greatest opportunities to advance open government include strengthening accountability mechanisms, expanding civic engagement with marginalized communities, and rebuilding trust in city institutions and leaders that have recently violated the principles of open government. The city government has taken a series of proactive steps to implement important infrastructure for advancing transparency and accountability. Leaders have a unique opportunity to create uniform standards for access to information, technological innovation, and democratic values of co-creations across all city institutions, especially those entrusted with the greater power and those benefiting from the greatest share of the city's resources.

The City of Buenos Aires developed and implemented its first action plan in a highly collaborative manner. The central axis of the commitments with greater progress were the opening of data in the three powers of the state and on sexual and reproductive health services. However, it would be necessary to advance in the prioritization of commitments that make a tangible difference in the lives of citizens, prioritizing inclusion and issues that, according to the diagnosis of the stakeholders, are part of the priority agenda of the city, such as security, habitat, housing, environment, social protection, gender, and access to justice. Finally, in the three cases, co-creating commitments and milestones that meet the public accountability dimension is a continuous challenge. Governments must include commitments that oblige government agencies to answer for the services provided and responsibilities acquired. Strengthening the assignment of rights, duties, and consequences for the actions carried out by officials and institutions could similarly broaden understanding and enforcement of the open government paradigm.

Notes

1. OGP launched the "Subnational Government Pilot Program" in 2016. This program consisted of 15 "pioneer" subnational governments who signed onto the Open Government Subnational Declaration and submitted their first Action Plans (APs) at the Paris Global OGP Summit in December 2016 which were implemented throughout 2017. Following the early results of the pilot period between 2016 and 2017, the OGP Steering Committee approved the expansion of the subnational pilot program, currently known as the OGP Local Program. (https://www.opengovpartnership.org/ogp-local-program/, accessed 3 November 2019).

2. For more information on the City of Buenos Aires' open government ecosystem please visit: https://gobiernoabierto.buenosaires.gob.ar (accessed 3 November 2019).

References

Abu-Shanab, E. A. (2015). Reengineering the open government concept: An empirical support for a proposed model. *Government Information Quarterly*, 32(4), 453–463.

Blanco, I., & Gomà, R. (2003). Gobiernos locales y redes participativas: retos e innovaciones. *Revista del CLAD Reforma y Democracia*, 26, 93.

Conradie, P., & Choenni, S. (2014). On the barriers for local government releasing open data. *Government Information Quarterly*, 31, 1(0), S10–S17.

Criado, J. I. (ed.) (2016). *Nuevas tendencias en la gestión pública: Innovación abierta, gobernanza inteligente y tecnologías sociales en unas administraciones públicas colaborativas.* Madrid, España: Instituto Nacional de Administración Pública.

Criado, J. I., & Ruvalcaba-Gomez, E. (2018). Perceptions of City Managers about Open Government Policies: Concepts, Development, and Implementation in the Local Level of Government in Spain. *International Journal of Electronic Government Research (IJEGR)*, 14(1), 1–22. ISSN: 1548-3886. DOI: 10.4018/IJEGR.2018010101

Criado, J. I., Ruvalcaba-Gómez, E., & Valenzuela, R. (2018). Revisiting the Open Government Phenomenon. *A Meta-Analysis of the International Literature. JeDEM, eJournal of eDemocracy and Open Government*, 10 (1): 50–81. ISSN 2075-9517 DOI: https://doi.org/10.29379/jedem.v10i1.454

Dawes, S. S., & Helbig, N. (2010). Information strategies for open government: Challenges and prospects for deriving public value from government transparency. In *International Conference on Electronic Government* (pp. 50–60). Springer Berlin Heidelberg.

ECLAC. (2016). Gobierno abierto. Comisión Económica para América Latina y el Caribe. Recuperado de http://www.cepal.org/es/temas/gobierno-abierto

European Commission. (2016). Open government. Recuperado de https://ec.europa.eu/digital-single-market/en/open-government.

Ganapati, S., & Reddick, C. G. (2012). Open e-government in US state governments: Survey evidence from chief information officers. *Government Information Quarterly*, 29(2), 115–122.

Gascó, M. (2014). *Open government. Opportunities and challenges for public governance.* New York, USA: Springer.

Grimmelikhuijsen, S. G., & Feeney, M. K. (2017). Developing and Testing an Integrative Framework for Open Government Adoption in Local Governments. *Public Administration Review*, 77(4), 579–590.

IDB. (2016). Reform/modernization of the state. Effective, efficient and open governments for the region. BID's Website. Recuperado de http://www.iadb.org/en/sector/reform-modernizationof-the-state/overview,18347.html

Jaeger, P. T., & Bertot, J. C. (2010). Transparency and technological change: Ensuring equal and sustained public access to government information. *Government Information Quarterly*, 27(4), 371–376.

Janssen, M., Charalabidis, Y., & Zuiderwijk, A. (2012). Benefits, adoption barriers and myths of open data and open government. *Information Systems Management*, 29(4), 258–268.

Lathrop, D., & Ruma, L. (2010). *Open government: Collaboration, transparency, and participation in practice:* O'Reilly Media, Inc.

Lee, G., & Kwak, Y. H. (2012). An open government maturity model for social media-based public engagement. *Government Information Quarterly*, 29(4), 492–503.

Meijer, A. J., Curtin, D., & Hillebrandt, M. (2012). Open government: Connecting vision and voice. *International Review of Administrative Sciences*, 78(1), 10–29.

Noveck, B. S. (2015). *Smart citizens, smarter state: The technologies of expertise and the future of governing.* USA: Harvard University Press.

OAS. (2016). Gobierno Abierto. Website, sección del Departamento para la Gestión Pública Efectiva, Secretaría para el Fortalecimiento de la Democracia (SFD). Recuperado de http://www.oas.org/es/sap/dgpe/ACCESO/G_abierto.asp.

OECD. (2016). Open Government, OECD's Website. Recuperado de http://www.oecd.org/gov/open-government.htm

OGP. (2011). Open Government Declaration. Recuperado de http://www.opengov-partnership.org/about/open-government-declaration

Red Gealc. (2016). La Red Gealc y el Gobierno abierto. Recuperado de http://red-gealc.org/gobierno-abierto/contenido/5595/es/

Ruvalcaba-Gómez, E. (2016). Participación Ciudadana en la Era del Open Government. Una mirada desde la literatura científica. *Paakat: Revista de Tecnología y Sociedad*, 6 (11).

Ruvalcaba-Gómez, E. 2019. *Gobierno Abierto: Un análisis de su adopción en los Gobiernos Locales desde las Políticas Públicas. Instituto Nacional de Administración Pública.* Madrid.

Ruvalcaba-Gomez, E., Criado, J. I., & Gil-García, J. R. (2017). Public Managers' Perceptions about Open Government: A Factor Analysis of Concepts and Values. In Proceedings of the 18th Annual International Conference on Digital Government Research (pp. 566-567). ACM.

Sandoval-Almazan, R. (2011). The two door perspective: An assessment framework for open government, *eJournal of eDemocracy and Open Government*, 3(2): 166–181.

Sivarajah, U., Irani, Z., & Weerakkody, V. (2015). Evaluating the use and impact of Web 2.0 technologies in local government. *Government Information Quarterly*, 32(4), 473–487.

Weyandt, Raymond. (2018). Independent Reporting Mechanism (IRM): Austin Progress Report 2017. Open Government Partnership.

Wijnhoven, F., Ehrenhard, M., & Kuhn, J. (2015). Open government objectives and participation motivations. *Government Information Quarterly*, 32(1), 30–42.

Williamson, V., & Eisen N. (2016). *The impact of open government: Assessing the evidence.* USA: The Brookings Institution.

Wirtz, B. W., Piehler, R., Thomas, M. J., & Daiser, P. (2015). Resistance of Public Personnel to Open Government: A cognitive theory view of implementation barriers towards open government data. *Public Management Review*, 1–30.

Yu, H., & Robinson, D. G. (2012). The new ambiguity of open government. *UCLA Law Review Discourse*, 59, 178–208.

Part II

Homeland security and human rights, a questioned balance?

5 Ethical controversies about Lethal Autonomous Weapons Systems

Views of small South American states

Raúl Salgado Espinoza

Introduction

Technology has brought a great deal of benefits to humanity, but its use in the development and innovation of weapons has converted it into a great danger for all individuals, particularly civilians affected by war and conflict. The development of the atom-bomb, for example, did not reduce the threat of war between states and their people, nor has it stopped the rise of new wars. Its use in Hiroshima and Nagasaki during War World II made it responsible for the death of hundreds of thousands of innocent civilians, despite the existence of alternatives approaches to its surrender (Burr, 2017).

Military technology and innovation, which cannot distinguish fighters from civilians and cannot reduce human suffering, is at risk of being prohibited. However, new military technology has always been the instrument through which wars have been won, domination and subjugation established, and hegemonies constructed. In that context, the acquisition and development of new military technology for some small states, particularly for those situated in regions of international conflict and who are surrounded by aggressive and belligerent forces, such as Israel, might appear to be justifiable and vital. Notwithstanding this, how can the deaths of innocent civilians by the use of new military technology in conflicts be morally justified? Should this new military technology, such as robots, unmanned aerial vehicles (UAV), and other Lethal Autonomous Weapons Systems (LAWS) be regulated or banned? These are the questions that have engaged academics experts, as well as states since 2014 within the United Nations Headquarters in Geneva.

For the developed big states or great powers, such as the USA and Russia, adequate regulation of the innovation, development, and employment by states of military technology, such as unmanned vehicles, drones, and other lethal autonomous weapons systems is needed, as this technology could help to reduce civil casualties in conflicts, protect cultural buildings, and even perform certain roles of a soldier such as inform and select targets, among others[1].

In contrast, other states, particularly small countries and most of the Latin American (LA) states see the acquisition and development of new military technology, particularly, the LAWS, as unjustifiable. Hence, they support a prohibition of further development and acquisition of these technologies as the states' statements and reports of the High Contracting Parties and of the Groups of Governmental Experts on LAWS show.

This chapter draws on the positions of LA states on LAWS and has its special focus on Ecuador and Uruguay. Ecuador has directly participated in the discussions since 2014 and Uruguay has not yet presented its point of view on these matters, although both states have been traditionally supporters of multilateralism, international norms, International Human Rights Law (IHRL), and International Humanitarian Law (IHL) within which this matter has been discussed. In this context, the leading question for this analysis has firstly been what are the views of the small states, Ecuador and Uruguay, on the acquisition and use of such military technology? The reflection about the use of such technology in wars and other international conflicts and their direct effect on the contenders as well as on the civil society has stimulated the political and ethical debate regarding principles and basic norms to be taken into account in military conflict and wars such as distinction, proportionality, and responsibility as basic principles of *jus in bello*. Indeed, the Geneva Conventions of 1949 and their additional protocols, which have been signed and ratified by Ecuador and Uruguay, also includes the regulation of distinction, proportionality, and protection or responsibility.

The employment of the new technology, such as armed robots, drones, and other unmanned vehicles such as LAWS are not clearly regulated. Therefore, the other leading question that can help the reflection in this chapter is what are the views of Ecuador and Uruguay on the regulation of the use of LAWS or the complete ban? The focus on the views of these two small states could help to understand the meaning of having and using this technology from the point of view of their unjustifiability as, on the one hand, these two states are geographically situated in a zone free from international conflict. On the other hand, these small South American states have traditionally been states that support international norms, particularly international norms that protect human rights. Hence, these case studies can help to construct a position of South American small states on the problems behind the indiscriminate, unregulated use and misuse of such weapons.

The data used in this chapter are documents and audios of the meetings obtained from the official websites of the United Nations (UN) Geneva and New York, of ministries of foreign affairs, as well as interior and subordinated institutions. Moreover, the interpretations of semi-structured interviews with state officials from the political and military institutions, as well as the analysis of press releases from the central government and from the local media have helped to understand the extent of the development and use of such instruments in Ecuador and Uruguay, as well as in the whole LA region. At the same time, it aims to evidence the level of reflection and discussion about the ethics of the use of these systems at war. The chapter draws on the principles of *jus in bello* to interpret the ethical implications around the discussion of a prohibition or regulation of the development and use of LAWS by looking at the positions of the Latin American states, particularly Ecuador and Uruguay. It aims to complement the reflection of Roff (2015), who highlights the issues behind an unregulated use of autonomous weapons systems in wars (AWS) in the context of the principles of *jus ad bello*.

The chapter develops the argument that Latin American states tend to support the creation of an internationally binding norm that prohibits emergent full LAWS and further regulates already existing weapons systems with certain autonomous capabilities. Ecuador supports this position in concordance with its constitutional legislation, ethical principles, and due to the impossibility that emergent and existing LAWS comply with the principles of IHL. Uruguay's international politics is coherent with the LA position on LAWS and its lack of formal national position is related to the character of small states diplomatic politics.

In order to develop this investigation, the text is structured into three main parts. The first section presents a reflection about the ethical implications of using new technology at war in the context of the existing academic literature, particularly in the Latin American region. The second section describes a general Latin American perspective on the debates about lethal autonomous weapons. The third section aims to develop and present the views of the South American small states, Ecuador and Uruguay, on the ethical implications about developing and using these systems at war.

The ethics behind military technology and reflections on the principles of *jus in bello*

The discussion about the ethical implications of the emerging military technology and particularly the development of lethal autonomous weapons systems has reached, on one hand, various circles of experts and academics from North America and Western Europe and at a more reduced level from East Asia and Latin America. On the other, state representatives have joined experts to discuss the issues behind the field of new military technology as such instruments of war have not yet been regulated and raises concerns about their use and misuse.

Taking into consideration that wars are conducted between two belligerent parts, an aggressor and the victim, it can be said that military innovation and technology can be developed due to various purposes and can be applied in various types of battlefields as shown in Table 5.1. For a defensive purpose, the military innovation and development of new technology can be based on moral principles and obligations that a state has towards its population and the international community. A state may confront aggression by using force for two reasons: "to defend the local object

Table 5.1 Purposes of weapons innovation and their field of application

Purposes/field of application	Purpose 1	Purpose 2	Purpose 3
Ground	Defensive	Offensive	Defensive & Offensive
Sea	Defensive	Offensive	Defensive & Offensive
Air	Defensive	Offensive	Defensive & Offensive
Cyber space	Defensive	Offensive	Defensive & Offensive

Source: author's design

of aggression – i.e., the victim state and its people" and "those foundational values of the society of states" (O'Driscoll, 2008; 12).

These defensive purposes are inevitably linked to an offensive act by the aggressor. Hence, innovation and development of weapons of any kind can have an offensive purpose, despite the unmoral content of aggressive actions by belligerent states. Moreover, defensive weapons can also have the capability to be converted and employed as offensive weapons as shown in Table 5.1 and all of them could be used at least in four battlefields: ground, sea, air, and cyber space.

In the context of conflict in any of the four battlefields, the manual use of weapons by any operator or soldier and with any of the three purposes has been regulated within the International Humanitarian Law (IHL). Similarly, the established historical principles of the rules of war, particularly within the customary and positive rules contained in *jus in bello*, has already been analyzed (Walzer, 1992). Hence, the behavior of a fighter outside these rules could be considered as a war crime.

However, the changing circumstances and practices of war, whereby technology and military innovation is an important factor, put the belligerent actors in unclear situations and problems regarding the behavior according to the ethics of war (Frost, 2005). The situations are even more problematic when military innovation seeks to replace human agency as well as consciousness about the principles of distinction, proportionality, and responsibility. As shown in Table 5.2 and taking into account Frost's (2005) proposal of war as a social practice, we could assume that war actors, such as policy makers and operators or soldiers, are conscious of these three principles and will have notions of the regulations made within customary law and its institutionalization through The Geneva Conventions of August 12, 1949 and complementary protocols.

The issues emerge when machines are not only instruments of war, but they are transformed into actors of war. This means this kind of technology will not have a human operator in charge (on the loop) once it has been activated. Hence, we question if it will have the capability of making decisions about distinctions between the belligerent and civilians or friend and enemy; the application of violence sufficient to wage a war (proportionality) and the responsibility of the warrior regarding civilians, injured soldiers, cultural objects, and buildings, among others.

Table 5.2 Principles of just war as social practice and actors of war

Principles of jus in bello/actors	Distinction	Proportionality	Responsibility/ precaution
Policy makers	√	√	√
Operators (soldiers)	√	√	√
Machines (LAWS)	?	?	?

Source: author's design

According to the debates and the summary of the reports of experts (Federal Foreign Office of Germany, 2016), at present behind every kind of machine used in wars, there is always a human, an operator on the loop. In that context, machines are still instruments of war. However, advances in technology and military innovation have already been implemented with capabilities to autonomously react to enemy presence and aggression. Hence these systems work with defensive purposes. Moreover at present, military technology such as robotic systems already perform various roles of human soldiers in conflicts, as well as working with certain autonomy as the Patriot Missile systems and the Phalanx system that poses capabilities such as select and engage (Arkin, 2013; American Society of International Law, 2014).

It is within this development that ethical implications of the use of these systems arise, especially considering that in the context of war, attacks and bombardments to civilians and non-fighting groups should be avoided, and although these weapons, when being used, are still monitored by humans, these new apparatus (LAWS) are not capable of functioning in ways that can still be driven by emotions and consciousness, as even in the context of war, the principle of humanity could be present in soldiers. Hence how could these weapons envisage and take into consideration the just war principles, particularly in fighting within wars of national liberation whereby the means of war are unconventional (Frost 2005). In this context, who should be accountable for the violation of the principles of distinction, proportionality, and responsibility? The question is more difficult to be answered when technology is not yet fully developed and the real consequences of LAWS in wars have not yet been experienced.

Therefore, in the last six years two major visions have been positioned in the academic debate. On one hand, some academics are in favor of a ban on these weapons, hence to put a stop to further development of LAWS, as they do not comply with the principles of IHL and basic rights to life (Asaro, 2012; Melzer, 2013; Marsh, 2014; Roff, 2015; Heyns, 2016). On the other hand, some scholars support the controlled development and a further regulation towards an appropriate use of this military technology, as such regulations could help in terms of transparency, a state's control, appropriate use, and in the case of deployment in conflict zones, they might even save civilian life (Anderson and Waxman, 2013; Gutiérrez and Cervell, 2013; García, 2015, García Rico, 2016; Lewis, 2015; Titiriga, 2016; Arkin, 2013).

Table 5.3 The problem of international regulation on the basis of morals

Morals/regulation	Right/just vs.	Wrong/ unjust
Legal	√	?
vs.	?	√
illegal		

Source: author's design

The ethical problem of this debate is that regulations are based on morals and there is a diversity of morals, which are based on diverse beliefs and cultural principles. Therefore, some people could conceive a situation of life or death as morally acceptable whereas for others it might be seen as unacceptable. Hence, some actions considered as just and right for some could be established as illegal and something that is considered as wrong and unjust by others could be regulated as legal. Table 5.3 illustrates this difficulty when dealing with regulation on the basis of diversity of morals and the definitions of justice and injustice.

From a simple point of view, it does not seem to be conflictive to agree on the illegality of a violent action considered wrong and therefore unjust, or actions being right and just, therefore legal. However, a problem arises, on one hand, when a violent action considered as wrong and unjust is legalized. On the other hand, an action considered right and just could be regulated as illegal, as according to the circumstances of war and power relations. The use of violence in the name of peace has historically been justified even by the philosophers of the Western and Islamic world (Idris, 2019).

To complement this debate, we should remember that there are various types of new technology, which in general can be equipped to function as autonomous weapons such as the use of AUV or drones, robots, computer viruses, and the adaptation of artificial intelligence for the decision-making process in the acts of war. Latin American academic reflection and research has focused mainly on the development and the use of drones in conflict and the incoherence with the International Humanitarian Law when using such instruments (Sánchez, 2014; Calvo Rez-Reguerel, 2014; Gomez Isa, 2015; Arteaga Botello, 2016; García Rico 2016; López-Jacoiste, 2018). For Jordán (2013) such worries, reflections, and debates around the use of new military technology and semi- autonomous weapons, such as drones, has become more important in those countries where such technology has been developed and has already been employed, as in the case of the use of armed drones by the NATO troops in Afghanistan (Jordán, 2013).

In Latin America there is limited research and academic reflection on the development and issues of autonomous weapons systems and LAWS. The research focus of the region is mainly on the ethics and regulation of civil use of drones (Gomis & Falck, 2015; González Botija, 2018), its use in human security, the application on policing responsibilities in controlling drug trafficking, border surveillance (Cawly, 2014) and on a limited military use for controlling, patrolling, and gathering information of the national borders, illegal mining, and deforestation. However, Latin American political practitioners and state representatives of around 17 countries have participated in the reflections and debates of the NU in Geneva about the issues arising with the development and use of LAWS. In that context, the following section seeks to construct a Latin American position.

General considerations and views from Latin America about LAWS

The Latin American region has traditionally been a strong supporter of International Human Rights Law (IHRL) and IHL. Hence, states have signed the Geneva Conventions and most of them have signed and ratified the additional protocols. In line with this, nineteen Latin American States are signatories of the Convention on Certain Conventional Weapons (CCW), including the small South American states: Bolivia, Ecuador, Paraguay, and Bolivia. In this context, we could say that multilateralism is one characteristic of Latin American international politics. Hence, the debates and reflections on the advances and use of military technology, especially the LAWS, carried out within the (CCW) found evident support by the LA states.

Around seventeen LA states have participated in the meetings and reflection since the initial discussions in the informal meeting of experts in 2014 to the last meetings of the Group of Governmental Experts (GGE) in 2019. Mexico, Costa Rica, Guatemala, Nicaragua, El Salvador, Honduras, Panama, the Dominican Republic, and Cuba, and eight South American states participated in the yearly meetings and debates and attended at least one of the various meetings. Paraguay, Bolivia, Surinam, and Guyana are the states with no direct representation and participation in these discussions and these two last states have not signed the CCW either. As shown in the first reports of the meetings of 2014 to 2016 and their statements of the parties involved, the central issues highlighted by the majority of LA states, particularly, the small states, there was a lack of clarity regarding the definitions of autonomous and automated weapons, their technological advances and the real magnitude of the violence that might cause the employment of such new technology.

This uncertainty is related to various factors. Firstly, as can be inferred from Table 5.2, military machines can be of different sizes and can have grades of automatization, from very basic automatic functions to fully autonomous functions after being turned on. As shown by the experts, as yet no states possess full LAWS, but various kinds of sophisticated weapons with certain autonomous capabilities have already been used. Secondly, none of the LA states are geographically situated in international conflict zones, although they maintain very conflictive societies within their own national borders.

Thirdly, Latin American states, particularly small states, do not possess this kind of military technology, so they are unfamiliar with specificities of their technical development. Moreover, since these states neither develop nor possess this kind of military equipment and there is not a clear national and international regulation, it can be perceived as an international and domestic threat. Hence, moral, ethical, and legal factors are strong arguments for national and international regulation of this kind of military technology. In this context, legal issues behind the decision-making process of the LAWS with compliance of the principles of IHL, referred to in Table 5.2, and the ethical principles regarding human rights, dignity, and compassion in situations of war have been positioned in the Geneva debates as one of LA states' concerns that require regulation.

Reflections at a multilateral level within the informal meetings of experts between 2014 and 2016 clearly show the kind of issues highlighted by LA states. The compromise with the principles of IHRL and IHL, particularly with the strict application of principles of *jus in bello* (distinction, proportionality, and precaution), have been turned into the measuring stick for new military technology and some existing weapons systems with certain autonomous capabilities, as there is no evidence that such legal and ethical principles cannot be met by using such weapons.

Hence, LA states argue that the contents of the Marten Clause are the most appropriate principles to be applied, taking into consideration that in cases of non-regulated factors the natural principles of right to life comes into place. This means that if LAWS had not been regulated yet, it does not mean that it is legal to develop and use them, and if this technology can not comply with IHL, it should be prohibited in accordance with Article 36 of the Protocol I of the Geneva Conventions. Statements from Argentina, Cuba, Ecuador, Colombia, Chile, and Mexico, among others, show that there is a regional consensus on the strict regulations of the emergence of LAWS. LA states base this position also on ethical principles of humanity, such as human dignity, compassion, and respect to the right to life. This technology might be equipped with algorithms based on questionable datasets, but it enables weapons systems to make decisions about human life and death. In this context, from the discussions of 2016, it can be said that the region was positioning the view of a prohibition of full LAWS and the requirement of new regulations to be applied to existing military technology of this kind.

The creation of the GGE on LAWS for the meetings since 2017, on one hand, enabled the LA states to formalize this view and to tackle many other concerns that emerged in the years that followed. For example, the necessity of continuing the technological development for civil and pacific objectives was mentioned. The worries to torpedo useful technological innovation became obvious in this context, as behind the development of these weapons there is a parallel technological development for civil and domestic use. This could be employed in various aspects of civil life. Hence, LA states support the idea of taking advantage of such technology. This argument is coherent with the present acquisition and development of new technology such as drones for domestic and policing roles, whose ethical discussions and regulations are also discussed within the domestic politics (Gomis & Falck, 2015; González Botija, 2018).

On the other hand, the urgency for regulating existing weapons with some autonomous capabilities that do not comply with IHL was considered in the debates. Argentina, Chile, and Ecuador, for instance, called for a moratorium on the development of these and the emergent LAWS until arriving at a consensus on the regulation that such weapons should strictly adhere to the demands of IHL. In this context, LA states agreed with the proposal that all autonomous weapons systems need to have a meaningful human control. The adoption of this concept is linked to the requirement of the principles of responsibility, proportionality, and accountability in case of exceeding the use of violence. Machines and humans might not always function according to the objective of the programmers and autonomous functioning of machines can also have biased systems, as these are based on datasets inserted by humans and with certain objectives. Therefore, LA states support the maintenance of responsibility, control, and accountability for the actions by the operator as well as by the state.

In conclusion, LA states propose a definitive prohibition of this kind of weapons if no human is on the loop. For the other kind of existing weapons systems there are two tendencies. Some LA states support the prohibition of all weapons that do not comply with the principles of *jus in bello*. Others tend to be more flexible and are hoping to come to a consensus regulation and control by the state of the whole process of development and by the human/operator while being used.

International norms, LAWS and the view of Ecuador and Uruguay

Small states rely on international organizations and multilateral politics as a lever for their foreign policy and to make themselves be heard internationally (Hey, 2003). Therefore, compliance with international norms and the building of a sense of international responsibility are the basis for their international politics. Another important aspect of small states' international politics is their interest in functioning international organizations, although by being a member of an international organization, small states can fall into the trap of restriction and security (Goetschel, 1998). From this perspective, it could be said that the engagement of South American small states, particularly Ecuador and Uruguay, in the construction and strengthening of functioning international organizations and international norms make up one more important national as well as international objective of their politics (Salgado Espinoza 2017).

Both, Ecuador and Uruguay are situated within an international peace zone and maintain a friendly and cooperative relationship with their neighbor states, despite some recent impasses between them. Ecuador ended its border conflict in 1998 with its southern neighbor, Peru, and they have since maintained a peaceful "brotherly" relationship. With its northern neighbor Colombia, Ecuador has not had a border conflict since the 1850s. However, their borders are the center of crimes and social problems, such as drug trafficking, illegal mining, environmental degradation, prostitution, and human trafficking (Pontón, 2018), and it is also used as hiding places for insurgent forces. This resulted in a break of diplomatic relations with Colombia when it bombarded Ecuadorian soil (Angostura) in order to kill Raúl Reyes, the second military commander of the Colombian Revolutionary Armed Forces (FARC) in 2008.

Uruguay has as its neighbors Argentina and Brazil. Uruguay has not had a military international conflict since the beginning of the 1870s, when the War of the triple Alliance (Brazil, Argentina, and Uruguay) against Paraguay ended. Since the beginning of the 1990s Uruguay has had a close political as well as commercial relationship with both its neighbors within the framework of MERCOSUR.

Regarding the memberships and contribution to the creation and the support to International Law, IHRL, and IHL, Ecuador and Uruguay show the highest records in agreements, signatures, and confirmation of these international norms in the region. More importantly, these states are part of the group of states that contributed to the actualization and creation of the Geneva Conventions following World War II. Similarly, they are signatories of the CCW since its creation, within which the debates on LAWS arose in 2014.

However, Ecuador has directly participated in the discussions since 2014, highlighting the need of the continuation of the debates within more formal meetings with the aim of constructing an international binding norm on LAWS. Hence, in 2016 Ecuador suggested the formation of the working governmental groups for discussing and constructing such norms. In 2018 and 2019 as part of the GGE, Ecuador strengthened its initial position regarding this technology and expressed its interest in contributing to the process of regulating the LAWS. On the contrary, Uruguay has barely participated in the debates. Hence, no contributions have been made and no formal position on these matters has been presented, although in 2014 and 2019 it registered to participate in the debates.

While Ecuador has maintained its position, since 2014, that full LAWS require a regulation that prohibits the development and use of such arms, Uruguay has prioritized other fields of multilateral politics. For Ecuador, LAWS are not compatible with existing international norms, whereas Uruguay has not formulated a final position.

What is relevant is that for Ecuador as well as Uruguay, existing domestic law dictates an international politics that support IHRL, hence IHL. This position is coherent with most of the reasons presented by the other LA states on the prohibition of emergent full LAWS. Already during the meeting of experts from 2014 to 2016, Ecuador's concerns centered on the capability of existing and emergent LAWS to comply with the principles of IHL. The establishment of the GGE on LAWS has opened the possibility for Ecuador to formalize its position and further contribute to the construction of an international binding norm for these kinds of new weapons.

In summary, Ecuador's position can be explained by looking at three factors: legal, ethical, and structural. The legal factors are firstly related to the Ecuadorian constitutional law. The Constitution of 2008 in its Article 416, numeral two, states that Ecuador "advocates the peaceful solution of international controversies and conflicts and condemns the threat and use of force as a means of resolving." Hence, the development and use of any kind of sophisticated weapons is incoherent with this national regulation. This same Constitutional article in its numeral four establishes as another principle of Ecuadorian international relations that the state "promotes peace, universal disarmament, and condemns the development and use of weapons of mass destruction..." In that sense, the development and use of military technology and new weapons are contrary to its national principles. Moreover, this Constitution states as principle of Ecuadorian international relations the "respect to human rights" and is centered on the construction of a society that "in all its dimensions respects the dignity of the individual and of the communities."

Secondly, these principles are also directly anchored to international legislation, especially to IHRL and customary law, which in the case of human rights issues takes precedent over domestic law. In this context, the position of Ecuador to construct an international binding regulation to prohibit the development and use of LAWS is supported by the explanations of the experts that there is no guarantee that existing and emergent LAWS can comply with the principles of IHL, particularly with distinction, proportionality, and precaution.

Thirdly, the Ecuadorian position is based on ethical factors, such as respect to life and human dignity, which are also highlighted in the Ecuadorian statements. These principles are entwined with issues highlighted by the group of experts regarding the difficulty of LAWS to take into account the morals, virtues, and human values at the moment of deployment and humanly uncontrolled action. Hence, Ecuador argues that the morals and values that support the ethical principles of responsibility, regret, and compassion, which can play a fundamental role in the heat of the moment in war, as well as the principles of accountability, cannot be transferred to machines.

Finally, structural international factors within which Ecuador interacts politically and in security matters also play an important role at the moment of presenting this position. Technologically, Ecuador is a long way from being able to develop capabilities to construct such weapons. Even for civil and policing use, Ecuador has imported drones from the technologically advanced countries (Estrada, 2011). Also, a project started in 2011 on the construction of UAVs which was developed with the support of local technical universities had as its aims their adaptation for military exploration, training, and national border surveillance (Loaiza, 2014). Politically, Ecuador is committed to the hemispheric regimes for human rights and multilateralism in the context of international security. Hence, its multilateral politics seek to find a coherence with the peers of Latin America.

Uruguay's lack of formal position regarding LAWS can be explained by looking at factors that characterize a small state. Particularly, when looking at the composition of their diplomatic service, small states' numerical deficit in their diplomatic service forces them to select foreign and international organizations for their participation. Uruguay maintains indeed a permanent mission in Geneva and was registered in the list of participants in 2014 and 2019. However, its voice has not yet been heard. Despite this, Uruguayan commitment to human rights principles, IHL, the CCW, and the peaceful resolutions of conflicts as established in its national constitution infers that it might take a similar position to the other LA states. Moreover, looking at the Uruguayan constitutional principles of Article 6 regarding international treaties, Uruguay must propose that "all differences that emerge between the parties must be decided through arbitration and other peaceful means." Hence, the use of military forces, development and use of any kind of LAWS in conflicts would be contrary to Uruguayan norms.

Notwithstanding this, both states see in the development and innovation of military technology a parallel development of technology that could be used for socio-economic development, as well as for civil and domestic use. In this context, Uruguay has decided to employ a system of drones to invigilate its borders with the objective of combating drug trafficking and other crimes.

Conclusion

This chapter has shown that in the context of the debates on the regulation or prohibition of LAWS, LA states tend to support a prohibition of emerging full LAWS and have called for a revision of the existing weapons systems with certain autonomous functions in order to determine the compliance with the principles of IHL, especially with the principals of *jus in bello:* distinction, proportionality, and precaution. The incapability of the technical accomplishment of these principals may result in a further regulation of such weapons. In this context, Ecuador, among other states has called for a moratorium on the development and use of such weapons until states come to a consensus on an international binding regulation. Ecuador's position of prohibition of full LAWS is coherent with its domestic and international legislation as well as with the LA position. Uruguay's lack of a formal state position could be based on the characteristics of a small state and its relatively small diplomatic service. However, its principles of domestic and international politics are coherent with the position of LA on LAWS. Both states can envisage two sides of technological military development. The destructive side and its incoherence with IHL and ethical principles regarding humanity and human dignity and the dangers in the development, acquisition, and use of LAWS.

Finally, it can be said that technology and military strategies have grown hand in hand within an environment of constant competition for power. Therefore, strengthening international institutions and norms could reduce the threats of living together within an anarchic international system, within which the development, having and employing these systems would have assured self-destruction, an arms race and the danger of imposition and subjugation. In this context, the South American small states as underdeveloped political entities in these matters can be seen as in a very vulnerable position.

Note

1. See the statements of all participant states, including the working papers summited by the USA and Russia at https://www.unog.ch/__80256ee600585943.nsf/(httpPages)/5535b644c2ae8f28c1258433002bbf14?OpenDocument&ExpandSection=3%2C1%2C2%2C7#_Section3

References

American Society of International Law. (2014). *Panel on Autonomous Weaponry and Armed Conflict.* Retrieved from https://www.youtube.com/watch?v=duq3DtFJtWg

Anderson, K., & Waxman, M. (2013). Law and Ethics for Autonomous Weapon Systems Why a Ban Won't Work and How the Laws of War Can. *American University Washington College of Law.* 2013:11.

Arkin, Ronald. (2013). Lethal Autonomous Systems and the Plight of the Non-combatant. *AISB Quarterly, 136.* pp. 1–9.

Arteaga Botello, Nelson. (2016). Política de la verticalidad: drones, territorio y población en América Latina. *Región y sociedad*, 28(65), 263–292.

Asaro, Peter. (2012). On banning autonomous weapon systems: human rights, automation, and the dehumanization of lethal decision-making. *International Review of the Red Cross*, 94, pp 687–709. Doi:10.1017/S1816383112000768

Burr, William (ed.). (2017). *The Atomic Bomb and The End of World War II: A Collection of Primary Sources.* National Security Archive. Documents 39A-B Magic. Retrieved from https://nsarchive2.gwu.edu/nukevault/ebb525-The-Atomic-Bomb-and-the-End-of-World-War-II/#_tc4

Calvo Gonzàlez-Regueral, Carlos. (2014). *Ética y legalidad en el empleo de drones.* Instituto Español de Estudios Estratégicos. Documento de opinión. Retrieved from http://www.ieee.es/Galerias/fichero/docs_opinion/2014/DIEEEO101-2014_Etica-Legalidad-Drones_CarlosCalvo.pdf

Cawley, Marguerite. (2014, April 18). Drone Use in Latin America: Dangers and Opportunities. *InSight Crime.* Retrieved from https://www.insightcrime.org/news/analysis/drone-use-in-latin-america-dangers-and-opportunities/

Estrada, Isabel. (2011). Vehículos aéreos no tripulados ayudan a la armada ecuatoriana a atrapar narcotraficantes. *Diálogos: revista militar digital*. Retrieved from https://dialogo-americas.com/es/articles/vehiculos-aereos-no-tripulados-ayudan-la-armada-ecuatoriana-atrapar-narcotraficantes

Federal Foreign Office of Germany. (2016). *Lethal Autonomous Weapons Systems: Technology, Definition, Ethics, Law and Security*. Retrieved from https://www.unog.ch/80256EDD006B8954/(httpAssets)/66E6253E1FECD90F-C12580A5002E879F/$file/GER+MFA+LAWS-experts+publication.pdf

Frost, Mervyn. (2005). Ética y guerra: más allá de la teoría de la guerra justa. *Revista Académica de Relaciones Internacionales*. 3: 1–27.

García, Denise. (2015). *Battle Bots: How the World Should Prepare Itself for Robotic Warfare. Foreign Affairs*. Council of Foreign Relations.

García Rico, Mar. (2016). Altas tecnologías, conflictos armados y seguridad humana. *Revista Iberoamericana de Filosofía, Política y Humanidades*, 36: 265–293. Doi: 10.12795/araucaria.2016.i36.12

Goetschel, Laurent. (1998). The Foreign and Security Policy Interests of Small States in Today's Europe. In *Small States Inside and Outside the European Union: Interests and Policies*. Laurent Goetschell (ed.). Dordrecht: Kluwer Academic Publishers; 9–32.

Gómez Isa, F. (2015). Los ataques armados con drones en derecho internacional. *Revista Española De Derecho Internacional*, *67(1)*, 61–92. Retrieved March 4, 2020, from www.jstor.org/stable/26180642

Gomis-Balestreri, M., and Falck, F. (2015). De ficción a realidad: drones y seguridad ciudadana en América Latina. *Ciencia y Poder Aéreo*, 10(1), 71–84. https://doi.org/10.18667/cienciaypoderaereo.430

González Botija, Fernando. (2018). Drones, seguridad pública y régimen sancionador. *Revista Vasca de Administración Pública*, 111, 271–310.

Gutièrrez, C, & Cervell, M. (2013). Sistemas de armas autónomas, drones y derecho internacional. *Revista del Instituto Español de Estudios Estratégicos*. 2013: 2.

Hey, Jeanne. (2003). *Small States in World Politics*. Lynne Rienner: Boulder.

Heyns, Christof. (2016). A Human Rights Perspective on Autonomous Weapons in Armed Conflict: The Rights to Life and Dignity. In Federal Foreign Office, *Lethal Autonomous Weapons Systems Technology, Definition, Ethics, Law & Security*. (pp. 148–160). Alemania.

Idris, Murad. (2019). *War for Peace: Genealogies of a Violent Ideal in Western and Islamic Thought*. New York: Oxford University Press.

Jordán, Javier. (2013). La campaña de ataques con Drones contra Al Qaeda en Afganistán. *Inteligencia y Seguridad: revista de análisis y prospectiva*, 14: 73–102.

Lewis, John. (2015). The Case for Regulating Fully Autonomous Weapons. *Yale L.J, 124*. Available at SSRN: https://ssrn.com/abstract=2528370

Loaiza, Carla. (2014) Las aeronaves no tripuladas ya se hacen en Ecuador. *Defensa: Revista del Ministerior de Defensa Nacional*, 1: 24–25.

López-Jacoiste, Eugenia. (2018). *Drones armados y el derecho internacional humanitario*. Instituto Español de Estudios Estratégicos. Documento de trabajo. Retrieved from http://www.ieee.es/Galerias/fichero/docs_investig/2018/DIEEEINV10-2018_Drones_DchoInt_Lopez-Jacoiste.pdf

Marsh, Nicholas. (2014). Defining the Scope of Autonomy Issues for the Campaign to Stop Killer Robots. *Peace Research Institute Oslo. Policy brief*. Retrieved from https://www.prio.org/utility/DownloadFile.ashx?id=128&type=publicationfile

Melzer, Nils. (2013). *Human rights implications of the usage of drones and unmanned robots in warfare*. European Union. Doi: 10.2861/213.

O'Driscoll, Cian. (2008). *Renegotiation of The Just War Tradition and The Right to War in The Twenty-First Century*. New York: Palgrave Macmillan.

Pontón, Daniel. (2018). *Drogas, globalización y castigo: Una aproximación a la gobernanza policial contra las drogas en Ecuador 2011-2016*. Cuyo. Tesis doctoral. Facultad de Ciencias Políticas y Sociales. Universidad Nacional de Cuyo.

Roff, Heather. (2015). Lethal Autonomous Weapons and Just ad Bellum Proportionality. *Case Western Reserve Journal of International Law*, 47 (1): 37–52.

Salgado, Raul. (2017). *Small Builds Big: how Ecuador and Uruguay contributed to the construction of UNASUR*. Quito: FLACSO-Ecuador.

Sánchez, W. Alejandro. (2014). *COHA Report: Drones in Latin America*. Retrieved from http://www.coha.org/wp-content/uploads/2014/01/COHA_Sanchez_LATAM_Drones_Final_Jan122014.pdf

Titiriga, Remus. (2016). Autonomy of Military Robots: Assessing the Technical and Legal ('Jus in Bello'). *The John Marshall Journal of Information Technology & Privacy Law*. 32: 2, pp 57–88. SSRN: https://ssrn.com/abstract=2602160 or http://dx.doi.org/10.2139/ssrn.2602160

Walzer, Michael. (1992). *Just and Unjust Wars: A Moral Argument with Historical Illustrations*. New York: Second edition: BasicBooks.

6 From sensationalist media to the narcocorrido

Drones, sovereignty, and exception along the US-Mexican Border

David S. Dalton

Introduction

On August 20, 2017, *The Washington Times* published the article "Drones Become the Latest Tool Drug Cartels Use to Smuggle Drugs into the U.S." According to the reporter, Stephen Dinan (2017), Border Patrol agents saw an "unmanned aerial vehicle (UAV)," which they followed to its landing site. Once there, they apprehended the stash, the drone, and even the person tasked with recovering the drugs. The article's headline implicitly proclaimed the advent of a new era of drug trafficking, but the article itself told a much less dire story. Drones represented a new challenge in an ever-evolving cat-and-mouse game between illicit traffickers and law enforcement, but overall drug imports to the United States had not changed significantly. Far from a game changer, drones simply provided a new means for smuggling contraband that had already been finding ways into the country for decades (Bunker and Mendoza, 2018; Fiegel, 2018, pp. 388–90). Even so, the mere idea that cartels could use drones played directly into a popular imaginary, in which transnational smugglers endangered US sovereignty. Far from harmless, this perceived threat provided an excuse for the United States to continue its own controversial drone activities, both along the border region and deep within Mexican airspace. In this chapter, I engage examples of cultural production from both countries to argue that drone activity along the US-Mexican border reflects and (re)produces an imaginary of the borderlands as an exceptional space where lawful institutions prove inadequate to maintaining order, and extralegal violence thus becomes necessary for maintaining sovereignty.[1] I begin with a discussion of how the specter of drug trafficking provides justifications for US drone surveillance both along the border and deep within Mexican territory, and I finish with a discussion about how narcocorridos like "Los Drones," by the band Los Alegres del Barranco, challenge a biopolitical context in which the Mexican government has ceded many of its sovereign responsibilities to the United States in the name of its so-called War on Drugs.

Drones and biopolitics along the border

Giorgio Agamben (2005) defines the state of exception as a "no-man's land between public law and political fact, and between the juridical order and life" (p. 1). Physical borders between nation-states occupy this no-man's land because "they find themselves in the paradoxical position of being juridical measures that cannot be understood in legal terms, and the state of exception comes in the legal form of that which cannot have legal form" (p. 1). The conditions surrounding the US-Mexican border region – particularly the fact that this invisible, arbitrary line separates the so-called Global North from the Global South – imbue it with an especial urgency. The natural result of these conditions is the clear distinction between what Agamben (1998) calls "bios" and "zoê" (pp. 9–12), two terms that the theorist uses to track the extent to which biopolitical societies value the lives of different people living within their borders. Bios refers to the "good" lives led by those who enjoy legal and political citizenship and protections, while the zoê lead what Agamben calls "bare life," a fact that bars them from full political and economic participation in society (Agamben, 1998, pp. 9–12). Viewed in this light, migrants become homo-sacer subjects: people who "can be killed and yet not sacrificed" because those in power do not view their lives as fully human (Agamben, 1998, p. 12). The biopolitical marginalization of immigrants plays out in an especially problematic fashion when we consider policies that place migrant lives in danger for the sake of border security.

Drone surveillance further exacerbates biopolitical divides along the border because it reduces the possibility for human interaction between Border Patrol and potential immigrants. It is in part for this reason that Mark P. Worrell (2015) asserts that UAVs in domestic space "keep American's [sic] divided between Us and Them and [provide] a symbol of external war within the fabric of home life" (p. 232). Drone technology thus exacerbates the state of exception and imposes an imperial biopolitical status quo that marginalizes people of Mexican descent throughout the country. Numerous scholars have argued that drone technology can lead to especially brutal forms of colonialist violence, because it removes the human element between aggressor and victim (Grossman, 1995, p. 102; see also Dalton, 2016, p. 20).[2] Of course, most research on drones – particularly its effects on drone pilots – focuses on its use in the warzones of the ever-expanding theaters of the War on Terror.[3] Nevertheless, studies on drones in those countries provide important insight into the technology's transposition to the US-Mexico border. Abraham Acosta (2014) notes the almost war-like state of the border region when he asserts that, "far from signaling the weakest point at which this biopolitical frame breaks down, it is in fact its most pronounced focus of power" (p. 224). Indeed, if we follow Agamben (1998), then it becomes clear that borderlands drone surveillance reflects the US-Mexico border's position as a "threshold [...] at the place of sovereignty" (p. 22). The biopolitical ramifications of these activities are thus part and parcel of a strategy aimed at projecting US sovereignty and influence both at home and abroad.

Certainly, UAVs interface differently with the reigning biopolitics when flown in domestic space. Ole B. Jensen (2016) alludes to this fact when he views the recent popularity of drones in the United States through the lens of Foucault's "boomerang effect," where technologies perfected in foreign wars are eventually adapted for civilian use (p. 20; see also Foucault, 2003, p. 103). Jensen (2016) further asserts that "the 'return' of drones in a strict technical sense is already a reality, but the regulatory frameworks and the underpinning dimensions of 'governmentality' are still open and contested terrains" (p. 21). While current laws inhibit drones from firing missiles in national territory, for example, Kyle Stelmack (2015) notes that police have been known to use these aircraft to shoot pepper spray and other nonlethal weapons at people in the country (p. 276). What is more, Stelmack (2015) worries that the legal framework for justifying the further weaponization of domestic UAVs is already in the works. His concerns take on greater significance as we consider the border region. As the threshold between the foreign and the domestic, this would be the logical site for the incremental weaponization of domestic drones in the future.

Drone activities emphasize biopolitical divisions between the United States and Mexico by collecting intelligence on foreigners – and even certain US citizens whom it deems as threats – in ways that challenge constitutional guarantees of liberty and privacy (Wolf, 2013). The principal goal of these UAVs, after all, is to find contraband – particularly drugs – and immigrants that have entered the country illegally. Saving dehydrated migrants figures relatively low as an objective, because it has no strategic benefits and because those affected do not enjoy a lofty position within the biopolitical structure on the northern side of the border (Acosta, 2014, p. 220; Inda, 2006, p. 174). Reports of Border Patrol members dumping out water meant for migrants in the Arizona desert during the presidency of Barack Obama – a time in which 593 migrants died of exposure while crossing the border – show the degree to which the logic of migrant expendability sits at the heart of current border policies (Carroll, 2018).[4] The fact that U.S. policy casts Mexican and Latin-American migrants as zoê does not mean that the Border Patrol will not, at times, highlight its humanitarian successes. The agency frequently invokes humanitarian ideals when attempting to secure greater approval ratings. Acting Commissioner of US Customs and Border Protection Kevin K. McAleenan's (2017) "CBP to the Rescue: Protecting the Homeland, Saving Lives," appears on the CBP webpage, for example, and it highlights the steps Border Patrol have taken to save "illegal aliens," a term that ironically interpellates migrants into zoê even as it affords them a degree of human rights. Of course, McAleenan neglects to address the fact that the migrants saved by his agency have to brave brutal desert conditions precisely because US border policies have driven them away from safer checkpoints (Rubio-Goldsmith et al., 2006).

It would be patently false to claim that U.S. policy aims to proactively kill migrants or even drug traffickers. Rather, it aims to push potential migrants to more dangerous routes – often unbeknownst to the migrants themselves (Acosta, 2014, p. 220) – in order to deter undocumented immigration into the country. Policymakers knew that their laws would result in more deaths, but they were willing to tolerate this in order to defend what they viewed as the integrity of the nation-state. The existence of narcodrones is merely one of the more recent developments to further exacerbate the border's exceptional status; it is probable that another technology or threat will come to replace drones as the new existential threat once these lose their allure. Indeed, the constant that appears across most discussions of so-called border security in the United States centers on drug trafficking, human trafficking, other forms of smuggling, and undocumented labor. As Walter Benjamin (1986) explains, "the tradition of the oppressed teaches us that the 'state of emergency' in which we live is not the exception but the rule" (682). While he does not cite Benjamin directly, Oswaldo Zavala (2018) recognizes this logic in Mexico and the United States, where he claims that drug cartels (and, one could argue, narcodrones) are the inventions of the very governments that combat them. As he argues, these states need a permanent enemy "that allows them to justify actions that would otherwise be illegal and even immoral" ["que permite justificar acciones que de otro modo resultarían ilegales e incluso inmorales"] (2018, p. 16; see also Mercille, 2011, p. 1649).[5] The threat – both real and imagined – of narcodrones ultimately serves to justify the continued militarization of the border.

The policy implications of this institutionalized state of exception are especially worrisome. Only a few years after Ronald Reagan declared the war on drugs on October 14, 1982, for example, Waltraud Queiser Morales (1989) proclaimed that "the 'evil empire of drugs' has the potential to evoke that fear of the enemy so basic and powerful in the doctrine of anticommunism" (p. 167). In the background of the assertion, of course, lies the argument that antidrug politics served to justify imperialist acts that the United States could not carry out under normal (read: non-exceptional) circumstances. This mentality appears in Aaron R. Schmersahl's (2018) MA thesis, *Fifty Feet Above the Wall: Cartel Drones In the U.S.-Mexico Border Zone Airspace, and What to Do About Them*, which he wrote for a degree in Security Studies from the Naval Postgraduate School. Here he argues that simply monitoring the border and US airspace is insufficient to defending US sovereignty. Rather, he asserts that Department of Defense drones should penetrate into Mexico to monitor "rogue drones" (p. 14).[6] Schmersahl's suggestion is problematic for at least two reasons: firstly, it proposes that the United States violate other countries' air sovereignty in order to defend its own against non-state actors; secondly, it calls for constant vigilance without providing measurable objectives to determine success. Schmersahl thus promotes an institutionalized state of emergency that will harm the most fragile sectors of Mexican society much more than it will any drug cartels.

Suggestions like this have led many people within Mexico to view American UAVs not as a means for protecting US sovereignty but as a new way through which the United States asserts its hegemony by interfering in what should be internal affairs. This is not to say that narcodrones are an entirely imaginary threat; they have been used for violence on a few rare occasions. On October 20, 2017, for example, police arrested four men in the state of Guanajuato who had fitted a plastic explosive and a remote detonator to a drone (Bunker and Sullivan, 2018). Following this isolated, crude attempt to use a UAV in a botched assassination attempt, numerous outlets both in the media and the academic community began to debate how to best confront the new reality of weaponized narcodrones (Associated Press, 2017; AM Noticias, 2017; Infobae, 2017).[7] These discussions played to fears within society at large, but they also overstated the threat that these drones represented. Indeed, as the hit confirmed, low-tech firearms – which were already prevalent – continued to be more effective than UAVs. Far from the great threat that some media and academic publications proclaimed, this attack showed that narcodrones remained in their infancy. Nevertheless, attacks like this also contributed to the perceived security threats that US officials depended on to further justify their own drone presence in the country.

Given this backdrop, it makes sense that many Mexicans would view the alliance between the United States and their own federal government – and not organized crime – as the principal villain in the drug-war crisis. Indeed, many Mexican thinkers assert that their country constantly accedes to the imperial drive of their northern neighbor. Jorge Volpi (2009) exposes the role of the United States and the Mexican government in exacerbating the drug conflict when he asserts:

> Drug trafficking is a criminal activity like none other: its existence depends on a public health decision imposed especially by the United States. The act of constraining individual liberties for moral reasons encourages the creation of providers who will take charge in administering these substances without caring about the pain that this brings about. The criminalization of drugs does nothing more than elevate their price, which grows proportionally to the public resources that are injected to fight the cartels.
>
> [El tráfico de drogas no es una actividad criminal como cualquier otra: su existencia depende de una decisión de salud pública impulsada especialmente por Estados Unidos. El constreñimiento de la libertad individual por razones morales alienta el surgimiento de proveedores que se encargarán de suministrar estas sustancias sin importar las penas que haya de por medio. La criminalización de las drogas no hace sino elevar su precio, el cual se incrementará en la misma escala en que se inyecten recursos públicos para luchar contra los cárteles.].
> (p. 129)

While Volpi focuses the majority of his critique on the United States, he implicitly upbraids the Mexican state for supporting a strategy that has both failed on its promise to curtail the illicit narcotics trade and exacted an unacceptable human toll on the people of Mexico. This perpetual violence further enriches Mexican politicians who sign lucrative military deals with the United States with the expectation that they will continue to prosecute a failed war.

The irony becomes especially cruel as we consider Luis Astorga's (2007) assertion that, while annoying, drug cartels have never represented an existential threat to the Mexican sovereignty (p. 54).[8] Many would argue that these claims reflect a reality that ceased to exist after Felipe Calderón declared war on drug cartels in 2007 (Sullivan, 2013, p. 175). Nevertheless, we should recognize the fact that Calderón's actions resulted from significant US pressure. As such, the narrative that organized crime represents the lone – or even principal – threat to Mexican sovereignty becomes especially suspect. Indeed, while drug trafficking has not historically posed challenges to national sovereignty, one could convincingly argue that the Mexican state has ceded many rights and duties to a foreign power by allowing it to dictate the country's narcotics policies. There are few areas that demonstrate this position more clearly than the use of drone surveillance deep within Mexican national space. In a report for the Council on Hemispheric Affairs, for example, W. Alejandro Sánchez (2014) noted that the United States, and not Mexico, generally piloted these aircraft (p. 2). By deferring its law-enforcement responsibilities to the United States, one could plausibly argue that Mexico also ceded its sovereignty in a way that drug traffickers had not. Zavala (2018) argues that the resulting biopolitics – which marks drug traffickers as killable homines sacri – reflects "the direct and absolute action of the state to preserve its integrity" ["la acción directa y absoluta del Estado para preservar su integridad"] (p. 93). He does not mention that state officials only achieve this end as they give in to US pressure and sell away their nation's sovereignty, but this fact is implicit in his critique. Paradoxically, then, Mexican leaders maintain state power in the domestic sphere by ceding their sovereignty to another country.

"Los drones": Resisting American hegemony through narcocorridos

The backdrop of the Mexican government's forfeiture of its own sovereignty figures heavily in criticisms against the federal government in contemporary cultural production. Indeed, it plays a key role in "Los drones," a narcocorrido by Los Alegres del Barranco. The song reverberates with popular notions of drug traffickers as modern-day Robin Hood figures who stand against globalized financial interests meant to crush not only them but ordinary Mexicans as well. As it portrays an outlaw who bravely stands against foreign-made (and piloted) drones and a complicit Mexican military, the song asserts a degree of comradery between drug traffickers and ordinary Mexicans that the government cannot replicate. This communion comes not because most Mexicans approve of cartel actions but because they oppose US incursions into their country. This shared, nationalistic community makes sense in light of Astorga's (1995) observation that the failed drug policy negates the legitimacy of the law and lawmakers (p. 144). The state uses the threat of drug trafficking and narcodrones to construct its own response as a type of "legitimate violence" (Astorga, 1995, p. 135; see also Zavala, pp. 44-51). Such an approach alienates many Mexican citizens – even non-criminals – because, legitimate or not, policies of violence produce collateral damage. Given the political context, it should surprise no one that narcorridos have grown in popularity even as the alliance between the United States and the Mexican state has escalated its prosecution of the drug war. Indeed, this genre of music provides a powerful site from which to challenge not only the Mexican state, but also the institutionalized states of exception that have contributed to violence along the border and throughout Mexico.

Kristine Vanden Berghe (2019) argues that the transnational nature of narcocorridos has led to the creation of a "narco community" that stretches across borders (p. 53). As she explains, "it is as if the narcos and druglords of Latin America symbolically come together in music to confront their common enemy: the political elites and U.S. institutions that combat them" ["es como si allí [en la música] los narcos y capos de América Latina se encontraran simbólicamente para enfrentarse contra su enemigo común: las élites políticas y las instituciones estadounidenses que las combaten"] (ibid). I would extend her argument to point out that, beyond bringing narcos from different countries together under a shared mythology, this genre of music also reaches non-narco listeners throughout Mexico – and the Western Hemisphere – and gets them to celebrate the heroes' exploits as well.[9] Narcocorridos thus create an Andersonian "imagined community" that stands in opposition to a federal state that capitulates to foreign interests (see Anderson, 1991). The subversive potential of this genre of music has led the Mexican government to employ numerous (mostly failed) strategies to suppress their diffusion.[10] The state's preoccupation with narcocorridos shows that these performances represent more than simply popular music: they challenge the country's antidrug policies by glorifying the exploits of drug traffickers while casting the federal government as corrupt and ineffectual. If we accept Zavala's (2018) argument that drug cartels provide a necessary foil for the Mexican governing class to justify its use of state power (p. 16), then it becomes clear that narcocorridos threaten to undermine that process by negating the governing class' claims to legitimacy.

The narcocorrido grew out of the traditional corrido, a musical genre that has challenged entrenched structures of power for decades. Some trace the genre back to the middle ages (Acosta, 2018, p. 817; Bergman, 2015), but the corrido grew into its own during the Mexican Revolution (1910–1917).

During that conflict, musical groups traversed the nation and proclaimed the victories of radical revolutionaries like Pancho Villa and (particularly) Emiliano Zapata through songs (Acosta, 2018, p. 818; Price, 2012, p. 53). While the narcorrido shares many similarities with traditional corridos, it also differs in several important ways. Mark Cameron Edberg (2004), for example, notes that narcorridos generally elide the long, historical exposition that characterized earlier corridos, opting instead to jump straight into the action (Edberg, 2004, p. 48). In this way they maintain their appeal to contemporary audiences who generally want more action and less context.

"Los Drones" begins somewhere in the mountains of the northern state of Durango during a chase between an unnamed narco, a drone, and Mexican soldiers. As the title suggests, the song challenges the institutionalized state of exception in which the United States habitually violates its southern neighbor's sovereignty – even if technically with the permission of the government – to spy on Mexican nationals. The song[11] begins in the following way:

> They chased me for five hours.
> First, they sent drones.
> They tried to hide the sound of the aircraft
> With the crowing of the roosters.
> They thought I wouldn't hear them,
> But they were wrong.
> I snatched up the R-15,
> But not the bulletproof vest.
> I'd gone to bed wearing my boots.
> I had no bike or truck,
> So I took off on foot.

The song paints a bleak picture not only of the narco's predicament, but of the state of exception that has militarized what most used to view as a law-enforcement issue. This song is especially interesting because it challenges the imaginary of technological sophistication among drug traffickers that US and Mexican media have promulgated through stories of narco-drones. Rather, this narco stands up to a disproportionately powerful and technologically advanced enemy. This element is key to the song's discourse; in taking on such a powerful enemy, the protagonist ensures his own glory, a key theme in almost all narcorridos (Acosta, 2018).

Rather than technological skills, the narco protagonist depends on his gut instincts; the song emphasizes this when he charges into the Durango mountains on foot. The song does not explain why he lost his vehicles, but the audience can imagine several possibilities. Perhaps the narco lost them in the attack, or perhaps he believes he will be harder to track if he traverses the sierra's harsh terrain on foot. That the song would so explicitly celebrate the narco's lack of technological sophistication is especially interesting given the Mexican state's emphasis on technological sophistication in its discourses of modernity (Dalton, 2018). The song provides a counternarrative to official discourses on modernity by showing a hypermasculine, heroic figure who eschews technology in order to stand against a more advanced foe. This should come as no surprise in light of Mabel Moraña's (2006) assertion that "many narratives articulated along the axis of violence represent conflicts and characters that evoke models of conduct and discursive ideals that would seem anachronic in the time where they are set" ["muchas narrativas articuladas al eje de la violencia representan conflictos y personajes que evocan modelos de conducta y discursividades que parecerían anacrónicas en los tiempos que corren"] (p. 187; see also Astorga, 1995, pp. 91–92). The nontechnological means through which the protagonist resists the drone-coordinated attack reverberates with her statement.

At the same time, the fact that the drones of the song lack missiles – or even pepper spray and teargas – suggests a more low-tech drone than what one would normally expect in a story proclaiming the terrifying power of military-grade UAVs. In this song, drones function primarily as tools for intelligence gathering and surveillance. Of course, this also reflects reality to some extent, given that drones in Mexican airspace tend to be used for surveillance rather than airstrikes. The song[12] captures this dynamic when several soldiers arrive to take the narco protagonist out:

> A boy and two artillerymen
> Got there right after I left.
> Only three of my men
> Went with me.
> "Boss they're coming after us,
> Give the order and we'll take them out."
> I told them it wasn't worth it
> To fight with those soldiers [*guachos*: bastards].
> But then I saw things weren't going
> The way that I expected.
> One of my buddies
> Took three bullets.
> Another buddy had a 50.
> When he realized what was going on,
> He rested his gun in the crotch of a tree.
> He made up his mind,
> He opened fire,
> And he chased the soldiers [*guachitos*: li'l bastards] away.
> That's when shit hit the fan.
> I fired on them with my R-15
> Those shot-up morons
> Retreated back to Culiacán.
> They chased me for 5 hours
> In the mountains of Durango.
> I'm telling you this, my friend [*pariente*: relative]:
> I need one more person

To join my crew.
I won't tell you my name;
Just know that if you join me
Things won't go too bad.

These verses mark a change in tone as the drone-induced violence becomes palpable. The idea that only three of the protagonist's men follow him suggests that the rest have either been arrested or, more likely, killed. The deaths of these men – presumably Mexican nationals – is the direct result of collusion between the Mexican state and a foreign entity. One of the most unnerving aspects of this song – and the narcocorrido genre of music in general – then, is that it recognizes and challenges the construction of drug traffickers as killable *homines sacri*. Beyond its celebration of violence, the song also denaturalizes the institutionalized state of exception that has marked drug traffickers like the charismatic protagonist as killable *zoê*.

"Los Drones" emphasizes that drone surveillance ultimately reflects (and facilitates) a dangerous state of exception that permits – and even demands – the extrajudicial killings of drug traffickers. At the same time, it would be absurd to code the lives that the protagonist and his men lead as "bare" because they have a great deal of money and artillery. The song reverberates with Alonso Salazar's (1996) characterization of narcoculture as a "premodern culture marked by consumption, raucous living, and death" ["cultura premoderna marcada por el consumo, la fiesta y la muerte"] (pp. 168–69). Certainly, Salazar directs his comments at narcoculture within Colombia. Nevertheless, he provides an interesting vantage point from which to view this narcocorrido. The protagonist's battle casts him in a heroic light precisely because he successfully engages and beats a more technologically advanced foe, and he achieves this by acting raucously with no fear of death. He thus becomes a symbol of resistance against both US violations of Mexican sovereignty on the one hand, and a complicit Mexican military and state on the other. The ambivalent nature of this song rings especially clear as we consider Mexican citizens' overall distrust of both the police and the military (Davis, 2006, pp. 55–56). "Los Drones" ultimately frames the violence of this particular conflict as one that the military brought on with its US-backed drone surveillance. Indeed, the narco and his men only engage after the military draws first blood. The song thus ironically asserts the narcos' innocence while at the same time accusing the US-Mexican antidrug alliance of indiscriminate violence.

"Los Drones" forcefully asserts common ground between drug traffickers, migrants, and everyday Mexicans. Each of these aforementioned groups ultimately lead lives as that powerful actors within the United States, and perhaps even within Mexico, coded as *zoê*. As such, any one of these people can be killed – or allowed to die – with impunity under exceptional circumstances. The firefight between the protagonist and the military thus represents a battle of necessity against an unjust biopolitics. The narcos' victory belongs not only to the protagonist but to any marginalized Mexicans who hear the song. Beyond criticizing constructs of Mexican – and particularly narco – bare life at a theoretical level, the narrator has also contested his zoê in practice. Similar to most narcorridos, "Los Drones" finishes when the narrator engages his audience. Unlike most narcocorridos, however, where the narco promises to harm those who would dare stand against him (Edberg, 2004, p. 48), this one ends with an invitation to join him. Because the narrator refuses to identify himself or the friend/relative to whom he directs his words, he becomes a synecdoche for the Mexican nation at large. If the people of Mexico stop tacitly supporting the US imperial project within their country and join the narco protagonist, then "things won't go too bad." The song thus implies that the United States and Mexico are dealing with a united front of drug smugglers – who, ironically, are more committed to maintaining Mexican sovereignty than are the country's elected leaders – and ordinary Mexican citizens.

Conclusions

This chapter has shed light on how drone activities along the US-Mexican border reflect, perpetuate, and even construct discourses surrounding the region's exceptional status. One key finding is that the advent of narco-drones does not represent the existential change that many academic and news sources would have us believe. Rather, these UAVs provide a new face for challenges that law enforcement has dealt with for decades. That being said, drones evoke ideas of technological sophistication that have further exacerbated the border region's exceptional status in the popular imaginaries of both the United States and Mexico by projecting the image that law enforcement is ill-equipped to contain the violent threats that exist in that supposedly lawless region. As such, officials in both Mexico and the United States capitalize on the mere possibility that drug traffickers could use UAVs to smuggle products to and from the United States. This threat, minor though it may be, provides the leadership of both nations with an institutionalized state of exception that allows them to undertake actions that would otherwise be deemed unethical or even illegal. This, of course, results from a polemical biopolitics that has both justified the further militarization of the border by the United States, and coerced Mexican officials into allowing US drones and personnel to operate within their national territory.

My comparison of US (and Mexican) reports of narcodrones has shown the extent to which key thinkers – both journalists and academics – have contributed to a discourse from which it is possible to institutionalize the border's exceptional state. The apocalyptic discussion of the potential of narcodrones to revolutionize drug trafficking implicitly validates strategic policies that would otherwise be unpalatable to large swathes of the population. Of course, many of the actions carried out by the US and Mexican governments have alienated people, particularly Mexican citizens who find these incursions offensive or even illegal. This fact has ironically opened a space for a shared communion between drug traffickers and Mexican citizens who have never been associated with organized crime; neither group approves when the government oversteps its authority or when it cedes its sovereign responsibilities to the United States. When the oppressed realize that their nation-state will not fight for their interests and needs, it is only natural that they would seek allies in other places, including organized crime. The discussion of "Los Drones" shows the depth of the imagined community that has come to exist between ordinary Mexicans – particularly migrants – and drug traffickers. In a high-tech world of drone surveillance and states of exception, it should come as no surprise that each of these groups would enjoy tales of anti-heroes who outsmart US drones and live to see another day.

Notes

1. The term imaginary here refers to the names and distinctions given to the border and to key actors within border politics. Significantly, this imaginary does not represent reality per se; rather it represents an imposition that we as subjects have imposed on that region. For a discussion of the tension between symbolic, imaginary, and real, see Slavoj Zizek (2006, 4).
2. Other scholars have challenged this narrative by pointing out that drone pilots necessarily understand the real-world effects of their actions. See Derek Gregory (2011, pp. 193-99) for an especially in-depth discussion on the videogame aesthetic of drone warfare.
3. See Louise Amoore (2006); Medea Benjamin (2013); Humeira Iqtidar (2016); Mark P. Worrell (2015).
4. The recent arrest of Scott Daniel Warren for aiding undocumented migrants further emphasizes the reach of these biopolitics, though the jury's refusal to convict him shows that certain segments of the U.S. legal apparatus continue to value Mexican lives (Associated Press 2019).
5. Unless otherwise indicated, all translations by the author.
6. Of course, the U.S. already pilots drones in Mexican airspace. See Alejandro Sánchez (2014).
7. Sullivan and Bunker's (2018a) anthology *The Rise of the Narcostate (Mafia States)* contains many academic discussions about drones and drug trafficking. Of special interest are Bunker and Sullivan (2018); Sullivan and Bunker (2018b); and Brenda Fiegel (2018).
8. Of course, Astorga (2007) does recognize that organized crime frequently challenged the sovereignty of state and local governments even prior to Calderón's war. See Astorga (2007, pp. 183-272).
9. It is for this very reason that Acosta (2015) argues that narcocorridos and narconarratives are complicit in normalizing and legitimizing drug violence.
10. For an in-depth discussion on the censorship of narcocorridos and the construction of a Mexican nation, see Hector Amaya (2013).
11. The Spanish version of the lyric can be found in the following site https://www.musixmatch.com/es/letras/Marca-MP/Los-Drones-En-vivo
12. The Spanish version of the lyric can be found in the following site https://www.musixmatch.com/pt/letras/Marca-MP/Los-Drones-En-vivo

References

Acosta, A. (2014). *Thresholds of illiteracy: Theory, Latin America, and the crisis of resistance.* New York: Fordham University Press.

Acosta, R. (2015). La narconarrativa: El papel de la novela y canción en la legitimación de los Grupos Armados Ilegales. In Oswaldo Estrada (ed.), *Senderos de violencia. Latinoamérica y sus narrativas armadas* (81–98). Valencia: Albatros.

Acosta, R. (2018). Taming heroes: Deep time, affect, and economies of honor and glory in contemporary Mexico. *Revista de Estudios Hispánicos* 52 (3): 815–36.

Agamben, G. (1998). *Homo sacer: Sovereign power and bare life* (D. Heller-Roazen, Trans.). Stanford: Stanford University Press.

Agamben, G. (2005). *State of exception* (K. Attell, Trans.). Chicago: University of Chicago Press.

AM Noticias. (2017, October 20). "Dron bomba" listo para detonar a distancia. Retrieved at https://www.am.com.mx/noticias/Dron-bomba-listo-para-detonar-a-distancia-20171020-0048.html

Amaya, H. (2013). Narcocorridos in Mexico and the new aesthetics of nation. In E. Thompson and J. Mittell (Eds.) *Media authorship* (506–24). Blackwell Press.

Amoore, L. (2006). Biometric borders: Governing mobilities in the war on terror. *Political Geography* 25: 336–51.

Anderson, B. (1991). *Imagined communities: Reflections on the origin and spread of nationalism.* London: Verso.

Associated Press (2017, October 24). Detienen a 4 con un dron equipado con una bomba. Retrieved at https://www.nbcwashington.com/news/local/detienen-a-cuatro-con-un-dron-equipado-con-una-bomba-violencia-guanajuato-452919433.html

Associated Press (2019, June 12). Jurors refuse to convict activist facing 20 years for helping migrants. Retrieved at https://www.theguardian.com/us-news/2019/jun/11/arizona-activist-migrant-water-scott-daniel-warren-verdict?CMP=fb_gu&utm_medium=Social&utm_source=Facebook&fbclid=IwAR1IK7qoOf2VnQ10dJh-br6Kq4fI7u3ClgDf4nFDExTTgBj43t3wmycck1xs#Echobox=1560356587.

Astorga, L. (1995). *Mitología del narcotraficante en México.* Universidad Nacional Autónoma de México.

Astorga, L. (2007). *Seguridad, traficantes y militares.* Mexico City: Tusquets.

Benjamin, M. (2013). *Drone warfare: Killing by remote control.* New York: Verso.

Benjamin, W. (1986). Theses on the philosophy of history (H. Zohn, Trans.). In H. Adams & L. Searle (Eds.), *Critical theory since 1965* (680–85). Tallahassee: Florida State University Press.

Bergman, T. L. L. (2015). "*Jácaras* and *narcocorridos* in context: What early modern Spain can tell us about today's narco-culture." *Romance Notes* 55(2): 241–52.

Bunker, R. & Mendoza, M. (2018). Mexican cartel tactical note #30: Marijuana kettlebells and catapult and air cannon projectiles. In J. P. Sullivan and R. J. Bunker (Eds.) *The rise of the narcostate (mafia states)* (281–93) [Kindle]. Retrieved from Amazon.com.

Bunker, R. & Sullivan, J. P. (2018). Mexican cartel tactical note #35. In J. P. Sullivan and R. J. Bunker (Eds.) *The rise of the narcostate (mafia states)* (542–53) [Kindle]. Retrieved from Amazon.com.

Carroll, R. (2018, January 17). US border patrol routinely sabotages water left for migrants, report says. Retrieved from https://www.theguardian.com/us-news/2018/jan/17/us-border-patrol-sabotage-aid-migrants-mexico-arizona

Dalton, D. S. (2016). Robo sacer: "Bare life" beyond the border in Alex Rivera's *Sleep Dealer. Hispanic Studies Review* 1: 15–29.

Dalton, D. S. (2018). *Mestizo modernity: Race, technology, and the body in postrevolutionary Mexico.* University of Florida Press.

Davis, D. E. (2006). Undermining the rule of law: Democratization and the dark side of police reform in Mexico. *Latin American Politics & Society* 48(1): 55–86.

Dinan, S. (2017, August 20). Drones become the latest tool drug cartels use to smuggle drugs into U.S. Retrieved from https://www.washingtontimes.com/news/2017/aug/20/mexican-drug-cartels-using-drones-to-smuggle-heroi/

Edberg, M. C. (2004). *El narcotraficante: Narcocorridos and the construction of a cultural persona on the U.S.-Mexico border.* Austin: University of Texas Press.

Fiegel, B. (2018). Narco-drones: A new way to transport drugs. In J. P. Sullivan and R. J. Bunker (Eds.) *The rise of the narcostate (mafia states)* (387-90) [Kindle]. Retrieved from Amazon.com.

Foucault, M. (2003). *"Society must be defended": Lectures at the College de France, 1975-1976.* New York: Picador.

Gregory, D. (2011). From a view to a kill: Drone and late modern war. *Theory, culture & society* 28(7–8): 188–215.

Grossman, D. (1995). *On killing: The psychological cost of learning to kill in war and society.* New York: Back Bay Books.

Inda, J. J. (2006). *Targeting immigrants: Government, technology, and ethics.* Oxford: Blackwell. Infobae. (2017, October 27) Drones armados con "papas bomba." Retrieved at https://www.infobae.com/america/mexico/2017/10/27/drones-armados-con-papas-bomba-el-nuevo-recurso-del-narco-mexicano/

Iqtidar, H. (2016). Conspiracy theory as political imaginary: Blackwater in Pakistan. *Political Studies* 64(1): 200–15.

Jensen, O.B. (2016). New "Foucauldian boomerangs": drones and urban surveillance. *Surveillance & Society* 14(1): 20–33.

Los Alegres del Barranco. (2017). "Los drones." Retrieved from https://www.youtube.com/watch?v=gfMLBlvrREQ.

McAleenan K. (2017, July 7). CBP to the rescue: Protecting the homeland, saving lives. *U.S. Customs and Border Protection.* Retrieved from https://www.cbp.gov/newsroom/blogs/cbp-rescue-protecting-homeland-saving-lives.

Mercille, J. (2011). Violent narco-cartels or US hegemony? The political economy of the "War On Drugs" in Mexico. *Third World Quarterly* 32(9): 1637–53.

Morales, W. Q. (1989) The War on Drugs: A new U.S. national security doctrine? *Third World Quarterly* 11(3): 147–69.

Moraña, M. (2006). Violencia en el deshielo: imaginarios latinoamericanos postnacionales después de la Guerra Fría. *Caravelle. Cahiers du monde hispanique et luso-brésilien* 86: 181–90.

Price, B. L. (2012). Where history ends and the corrido begins in Pedro Ángel Palou's *Zapata. Latin American Literary Review* 49(79): 45–60.

Rubio-Goldsmith, R., McCormick M. M., Martínez D., & Duarte, I.M. (2006). The "funnel effect" and recovered bodies of unauthorized migrants processed by the Pima County Office of the Medical Examiner, 1990-2005. (Report Submitted to the Pima County Board of Supervisors). Tucson: Binational Migration Institute, Mexican-American Studies and Research Center at the University of Arizona.

Salazar, A. (1996). *La génesis de los invisibles. Historias de la segunda fundación de Medellín.* Bogotá: Programa por la paz Compañía de Jesús.

Sánchez W. A. (2014, December 1). Council on Hemispheric Affairs Policy Memo #4: Drones in Latin America [PDF file]. Retrieved from http://www.coha. org/wp-content/uploads/2014/01/COHA_Sanchez_LATAM_Drones_Final_ Jan122014.pdf

Schmersahl, (2018). A. R. *Fifty feet above the wall: Cartel drones in the U.S.-Mexico border zone airspace, and what to do about them* (Unpublished masters thesis). Naval Postgraduate School, Monterey.

Stelmack, K. (2015). Weaponized police drones and their effect on police use of force. *Journal of Technology Law & Policy* 15: 276–92.

Sullivan, J. P. (2013). How illicit networks impact sovereignty. In Miklaucic, M. & Brewer, J. (Eds.) *Convergence: Illicit networks and national security in the age of globalization* (171–88). Washington: National Defense University Press.

Sullivan, J. P. and Bunker, R. J. (2018a) *The rise of the narcostate (mafia states)*. [Kindle]. Retrieved from Amazon.com.

Sullivan, J. P. and Bunker, R. J. (2018b) Narcodrones on the border and beyond. In J. P. Sullivan and R. J. Bunker (Eds.) *The rise of the narcostate (mafia states)* (38–45) [Kindle]. Retrieved from Amazon.com.

Vanden Berghe, K. (2019). *Narcos y sicarios en la ciudad letrada*. Valencia: Albatros.

Volpi, J. (2009). El insomnio de Bolívar. *Cuatro consideraciones intempestivas sobre América Latina en el siglo XXI*. Mexico City: Random House Modadori.

Wolf, N. (2013). The coming drone attack on America. *Peace and Freedom* 73(2): 8–22. Retrieved from https://search.proquest.com/docview/1790927735?pq-origsite =gscholar.

Worrell, M.P. (2015). Imperial homunculi: The speculative singularities of American hegemony (drones, suicide bombers, and rampage killers, or, an excursion into Durkheimian Geometry). *Globalization, Critique and Social Theory* 33: 219–41.

Zavala, O. (2018). *Los cárteles no existen*. Barcelona: Malpaso.

Zizek, S. (2006). *How to Read Lacan*. New York: W. W. Norton & Company.

7 The process of technologization of the drug war in Mexico

Avery Plaw, David Ramírez Plascencia, and Barbara Carvalho Gurgel

Introduction

Mexican narco gangs, despite considerable state efforts to destroy them, are growing and spreading. Narco gangs are present practically in all continents, consequently the drug war is no longer a restricted Mexican problem, but an international threat that has exacerbated crime and violence far beyond national borders. Gangs are well equipped, and they have the logistic of their trafficking routes, setting efficient supply routes, and finding inventive mechanism to recover their revenues. How are they flourishing despite local and international security strategies, and huge invesments in national safety? Existing research has suggested a number of factors, including growing markets at home and abroad (Durán-Martínez, 2015; Bergman, 2018), the increased poverty and desperation in countries where production occurs (Zepeda Gil, 2018) and the inefficiency of traditional strategies used to eradicate narco gangs (Vilalta, 2014; Lindo & Padilla-Romo, 2018). In this chapter, we will show that one major factor in their success is the rapid adoption and creative exploitation of cutting-edge technologies. In section I, we provide a general outline of the drug problem in Mexico; in section II, we focus on how information technologies shape transnational crime and in section III we focus on the analysis of the process of "technologization" of the drug war in Mexico. Fieldwork stands on the analysis of relevant cases covered by media online news, print press, television related to the use of information technologies, and the drug problem in Mexico. We pay particular attention to what kinds of technologies are used, but also to how this process of technologization has shaped the actions and strategies of the diverse actors involved: citizens, civil organizations, governments, and the narco cartels.

The drug problem in Mexico

There have been well organized transnational drug trafficking networks in Latin America since the 1970s, where numerous countries collaborate in one way or another (Kenney, 2003): from plant growing and the processing of narcotic drugs, to the final distribution of the product (Millard, 1997).

In the case of cocaine for example, the plant is cultivated in Colombia and Peru, then the product is treated using chemicals from Brazil, and later, the product travels to Mexico, where Mexican cartels bring it into the United States. After the shutdown of the cocaine route in Miami by the United States Coast Guard in the mid-eighties, Mexico was consolidated as the main route to export this drug to the United States (The Economist, 2011). Despite transnational efforts to decrease the Mexican cartels' influence, it continues to grow and spread.

Recently, Mexican cartels have taken benefit of the partial success of The Plan Colombia, a US strategy to reduce the power of Colombian drug cartels and decrease the frequency of the cocaine-smuggling deliveries to the United States (Bagley & Rosen, 2015). In some cities, they even have a near monopoly of the distribution of heroin and cocaine, as the Sinaloa Cartel has established in Chicago (Bates, 2014). Due to the key role that these crime organizations play in the narcotrafficking problem, US and Mexican authorities have undertaken important efforts to eradicate them. Traditionally, one of the most important strategies to undermine the power of the cartels, called the Kingpin strategy, is to capture or kill the leader of the organization (Kenney, 2003; Lindo & Padilla-Romo, 2018). It has frequently been used in Colombia and Mexico. However, the effectiveness of this measure is controversial, since often, after the death or imprisonment of the head of the organization, other people inside the organization assume the role of the leader, and consequently the criminal group remains undiminished. Indeed, this process of adjustment inside the group, rather than having the effect of reducing its power, sometimes allows for its modernization, since the fresh leaders learn from their predecessors and change the routes, the practices, and improve the logistics to avoid law enforcement.

In Mexico, the "kingpin strategy" has exacerbated the murder rate across the country in the last decade (Vilalta, 2014; Lindo & Padilla-Romo, 2018). After the death or neutralization of the kingpin, a struggle is triggered among the sub leaders inside the organization to gain control. In some cases, the cartel even splits into two or more groups, as in the case of the cartel *Nueva Generación* (New Generation), a group that emerged as an armed wing inside the Sinaloa Cartel, but which now is enduring a process of segregation that has started a war among criminal cells from the group in Guadalajara, Mexico, the headquarters of this organization (44Lab, 2018). This rising of violence among criminal gangs has forced the Mexican government to send the army to patrol on the streets. But even with a strong military presence on the streets, fighting the narco gangs has become a hazardous task. Criminal organizations are well armed, and they have improved their methods of bribing public officers, producing transnational trafficking routes, and establishing efficient local distribution networks, as well as finding innovative ways of recovering their money through international money laundry schemes (Bergman, 2018). In addition, thanks to globalization and the development and improvement of financial systems, the drug problem is no longer a Mexican or Colombian issue. It has become an international threat that has exacerbated crime and violence far beyond the national borders (BBC, 2008). Cartels have established business and distribution networks in every region in the world: from the United States to Europe and Asia, and even as far as Africa.

There are other important trends that shape the actual drug problem in Mexico, like the global spread of consumption, the proliferation of farming zones, the diversification of trafficking routes, and the fragmentation of crime organizations (Bagley & Rosen, 2015). This fragmentation in Mexico has caused a great amount of casualties, creating a more complex landscape and discouraging the appearance of an integral solution to the problem, which focuses not just on the incarceration of the gang leaders, but bringing peace and development to impoverished zones of the country, where people are forced to work for the organized crime organizations, like farmers growing poppies or marijuana, becoming drug dealers or even hitmen. In addition, the consumption of some drugs like cocaine have increased, especially in countries traditionally associated with its production and smuggling like Colombia and Mexico (Durán-Martínez, 2015). This internal demand has created novel local markets that make the business even more lucrative. Finally, another key factor that shapes the actual drug problem is the arrival and global spread of digital technologies: tablets, smartphones, social media and mobile applications, the socialization of novel technologies such as drones, and the use of bitcoins.

Information technologies and transnational crime

Since the 2001 terrorist attacks in New York, there has been an increased worry about security in North America and internationally (Bromby, 2006). This global security concern has been accompanied by the development of technological devices and software to support surveillance and prosecution of electronic crimes, the detection of possible terrorist activity, and other things of this nature. Electronic surveillance is becoming almost omnipresent in airports, at home, at school, and in offices (Selgelid, 2011; Engberts & Gillissen, 2016). Big data and artificial intelligence allows the development of sophisticated surveillance systems that not only are able to review large amounts of data in real time, but they can make decisions on how to proceed when facing an illegal activity without the intervention of humans (Hallevy, 2015; Castelli, Sormani, Trujillo, & Popovič, 2017). Crime organizations have been quick to adopt novel information technologies as well. They have incorporated digital media to improve diverse activities inside the organization: to consolidate and spread their operations across the globe, to recover their business profits through money-laundering systems, to improve the logistics of their shipments and the distribution of the product to the final consumers, and to monitor and get tactical information about the authorities and the activities of other cartels. They have even adopted emerging technologies like drones and other autonomous vehicles like submarines to carry thousands of pounds of drugs across national borders, or even to modify domestic drones, incorporating explosives to attack the authorities. This process of "technologization" inside the drug gangs presents an important challenge for local and international security offices. Authorities, for example, must face the problem of fighting cartels and criminal transnational organizations that have found shelter inside dark virtual networks (Millard, 1997; Angelini & Gibson, 2007). Inside these spaces, criminals and terrorists operate with a certain level of freedom. This lack of control is caused mainly by two important facts: (a) the fast development and sophistication of digital technologies like encryption that enables the existence of these unregulated spaces, in constant upgrade, allowing them to avoid police surveillance (Keene, 2011); and (b) the scarcity of a proper legal framework which would not only allow authorities to combat cybercrime but also to bring criminals before the courts. This is especially important when considering misconduct that is committed through an international network like the internet, where legal issues usually involve diverse and contradictory legal systems and jurisdictions. National and international authorities make important efforts to eradicate these zones, such as with the shutdown of The Silk Road marketplace, a darknet where clients and customers sell and acquire diverse illegal drugs. However, these spaces are in constant evolution, moving from one platform to another, taking advantage of novel applications and protocols that support anonymous communications across the Internet.

The process of "technologization" of the drug war in Mexico

More than fifteen years after the start of the war on drugs in 2006, the outcomes are not promising. The militarization of the operation against the criminal gangs has been ineffective in reducing the violence; on the contrary, it has increased the murder rate and triggered human rights violations and abuses by soldiers, particularly in poor and isolated communities (LaSusa, 2017). Moreover, thanks to the corruption and inefficiency of government agencies, criminal gangs have acquired powerful weapons to fight the local authorities and even the Mexican army. Their armaments and resources easily exceed the capacity of the county police officers, particularly in rural areas. Most of the time they are forced to cooperate with the criminal organizations, as occurred in the case of The Ayotzinapa in 2014, where dozens of students were captured by the local police, which were under control of the cartel *Guerreros Unidos* (United Warriors), and then delivered to the drug cartel which made them disappear or killed them. Due to the failure of local authorities to protect the lives and properties of citizens, soldiers and federal police have been required to guard several areas across the national territory. The situation has gotten worse following Donald Trump's arrival to the White House. Some topics such as "the Central American refugee crisis," "NAFTA," (now the "USMCA") and "The Wall" have now become relevant components in the debate about domestic security policies on both sides of the border. Trump has not been hesitant to accuse Mexican authorities of not doing enough to stop illegal immigration from Central America, and to eradicate the narco cartels. On the other hand, the current Mexican president, López Obrador, has not been successful in opening a suitable channel of communication with his US counterpart. Trump's accusations and threats, along with the profound challenges of managing the Central American migrant crisis, have obstructed mutual cooperation in recent for years. At a local level, factors like the lack of coordination among the local and federal Mexican authorities, poverty and social inequality, and the persistent public corruption (Zepeda Gil, 2018), have caused, besides high investments on equipment, the profesionalization of local forces, international collaboration amid Mexican and United States authorities, the militarization of the police and the investment in surveillance technologies; the narcotraffic problem is far from an end (The Economist, 2018). In 2019, at the beginning of Andrés Manuel López Obrador's presidency, there has been an increase of 9.6% in the murder rate in the first three months of 2019, in comparison with 2018 (BBC, 2019).

Another aspect that has shaped the evolution of the drug war in Mexico is the incorporation of digital technologies as strategic tools not only inside the criminal gangs, but in the government and civilians as well. In this equation, the use of social media and mobile devices has played a key role among communities to share information, improve logistics, and denounce criminality. Indeed, Mexico is one of the countries with the highest levels of social media penetration in the world. Almost all the internet users, about 80 million, have at least one social media profile (Asociación de Internet MX, 2019). The main source of information for Mexicans is not television or radio but the internet, which is mostly accessed through mobile devices, particularly smartphones. This fact is promising; there is potential to have a decentralized and fresh communication media that is available in a cheap way for millions of people, that could give people access to a vast source of information from their mobile devices or personal computers. However, this quick spread of digital technologies has triggered serious social problems as well, including cyberbullying, the invasion of privacy, online fraud, and the proliferation of fake news and bots which are used to set up disinformation campaigns to subvert the political and social order. The drug problem has not been immune to this trend (Muoio, 2016). Now, with the arrival of technologies like social media, smartphones, and drones, it has reached a new level. For decades, the members of drug gangs have been the subject of traditional media coverage, by television, radio, or newspaper. As such they have been the source of inspiration for the composition of hundreds of narcocorridos, a subgenre of the regional Mexican corrido (narrative ballad), which is centered on the life of drug smugglers. With this sympathetic treatment, there has not been a huge need inside the criminal gangs to contact the media and openly express their opinions or agenda to a wider public (ScienceDaily, 2014). In recent years, however, there have been a growing number of cases in which the narcogangs not only share multimedia material such as videos, photos or audio files with big and small media corporations, but also organize meetings with journalists and actors to hold interviews, as in the case of the famous cartel leader Chapo Guzman, who met with the Mexican actress Kate del Castillo, well known across Latin America for her performance in the tv show "La Reina del Sur" (The Queen of the South) and the actor Sean Penn. Another of the methods employed by the narcocartels to contact the public has been the use of "narco mantas" (Mendoza Rockwell, 2010), a message left by a drug cartel on a cloth banner, containing threats or explanations for their actions. Typically, the people or the local authorities find and report the message and then the media spreads the content (Cerda Pérez et al., 2013). In some cases, the "manta" has been left with corpses or human remains, usually a member of the opposing cartel or a police officer who was previously kidnapped. Frequently, the "narco manta" have been used to denounce the authorities, or to threaten a criminal group or an individual. On other occasions, the authors asked the public for understanding and empathy. As soon as social media became popular in the country, the internet was flooded with a huge amount of information related to the confrontations among the public forces which the narco gangs sometimes broadcast live. Most of the content published is related to executions, public statements, or threats to their enemies and competitors (Atuesta, 2017). In the beginning, the narcogangs did not directly publish the videos or photos. Later, when they did begin to post themselves, they did it through sites like "El blog del Narco" (Arsenault, 2011; Monroy-Hernández & Palacios, 2014), a virtual blog created in 2010, and which is still live. The information inside the blog is usually uncensored, and for that reason it is common to find pictures and videos of executions and murders.

Social platforms and the adoption of mobile devices allows a more diverse and multimedia coverage of criminal incidents, attacks, and narco messages. It is not only big media corporations that participate in this media coverage, but also small media outlets, and even the public, who are able to share information and data. As the drug war evolved and spread across the country, the internet has not only emerged as one of the most important news sources on criminal incidents, but it turned "the drug war" into a global media phenomenon, and even a massive hit on Netflix, with bio series on famous cartel leaders such as Pablo Escobar or Joaquín "El Chapo" Guzmán. In short, criminal organizations, despite the illegal nature of their activities, have come to occupy an important place in social media which they exploit to their advantage. They use these virtual spaces to recruit new members, to promote their criminal activities, and as a marketplace to sell their "product" (Nix, Smith, Petrocelli, Rojek, & Manjarrez, 2016). However, the media overexposure of narcotrafficking has also brought collateral damage to a critical point. The number of journalists kidnapped and killed has risen (Bustamante & Relly, 2014), as they usually work without protection and

Table 7.1 The narco cartels' use of information technologies

The Cartel	Technology	Use
Diverse Cartels (circa 2009)	narco mantas	To threaten other cartels or authorities. Sometimes to gain the favor of the communities where they have a presence.
Diverse Cartels (circa 2010)	Social media	To threaten other cartels or authorities. Sometimes to gain support in local communities or even in the general public (i.e. the demonstrations against the incarceration of El Chapo. In addition, to advertise the cartel and its products.
Diverse Cartels (2014)	The construction of a radio network	To use corporate facilities and infrastructure to create their own communication network
Unknown cartel (2015)	Drones	To transport drugs across the border
The Gulf Cartel or Zetas (2015)	Cell Antenna, Video cameras (118) and 59 illegal communication antennas	Conduct surveillance on the army in the state of Tamaulipas
Cartel CJNG (2017), Unknow Cartel (2018)	Drones with explosives (2017)/ grenades (2018).	To commit attacks against authorities or members of opposite cartels.
Diverse Cartels (2018)	Cryptocurrencies	The use of cryptocurrencies to launder illegal money by Mexican cartels.
Diverse Cartels (2019)	Smartphones and mobile applications.	The use of smartphone applications like WhatsApp to sell drugs, coordinate logistics and surveillance operations.

Source: Authors.

are vulnerable, since their investigations affect the narcos, the government, and corporate interests (Nicolás Gavilán, 2018). Yet social media is not the only technology that became a valuable asset to narco gangs (see table 7.1). They have recruited telecommunication engineers, even kidnapped them, to set radio networks to support the interchange of information among the members through the usage of walkie-talkies. The use of radio networks has given them a significant advantage compared to other recent media like smartphones since radio networks, unlike smartphones, are cheaper, less traceable, more secure, and it provides communication in remote areas out of the cellular phone coverage (Tabor, 2014; González, 2018).

Narco gangs, in Mexico as well as Colombia, have also used light aircraft, speedboats, and even submarines to bring drugs into the United States. But it was with the arrival of El Chapo that other inventive techniques have been added. El Chapo had been a specialist in constructing smuggling tunnels or "narco-tunnels" across the US-Mexican border, secret underground passages used for smuggling drugs, guns, and people (Schneider, 2019).

Another device used with more frequency among the gangs are drones, or unmanned aerial vehicle (UAV). Drones are commonly used for military purposes, particularly in executing dangerous combat missions where the pilot's life would otherwise be at risk, such as in the Syrian War. The advantage of using drones as carriers for illegal substances is significant. It is easier and less expensive than building a tunnel or operating a submarine. For the price of one drug submarine, about one or two million US dollars (Woody, 2019), it is possible to buy dozens of aerial drones capable of carrying product, which increases the chances of sending more drugs across the border. Every UAV is capable of transporting about one hundred pounds and cannot be detected by radar as light aircraft can, which diminishes the chance of being captured by the border patrol (Green, 2014). Since the incorporation of drones among the Mexican narco gangs in 2015, this technology has been used to surveil the movement of authorities and rival groups, and carry drugs across the US-Mexican border, but some recent cases in the city of Salamanca in the state of Guanajuato in 2017 (Axe, 2017) and Tecate, in the state of Baja California (Sullivan, Bunker, & Kuhn, 2018), show how criminal gangs have started using these devices to attack their rivals or members of the public force. In the Tecate case, the target was the Baja California Public Safety Secretary, Gerardo Sosa Olachea. Two drones were used, one of them was armed with two grenades that did not explode. In addition, narco gangs have started employing the services of the Chinese underground banking systems (CUBS), as a way to evade financial controls and to "legalize" across Bitcoins the revenues of their businesses (Emem, 2018).

Following this technological trend, Mexican authorities, civic organizations, and communities have gradually incorporated digital technologies in the fight against criminal gangs (See Table 7.2). Throughout the years, social platforms like Twitter or Facebook have become ideal spaces to denounce criminality and provide strategic information to support the capture of offenders (Calderón, 2012). In addition, other special digital services have been developed which support communication among the citizens and the police. The first services of this kind allowed people to send text messages using their cellular phone, or to post the information directly to a website. This information allowed the authorities to have a more accurate map of the crimes and the geographic zones where they occur (Piña & Ramírez, 2019). Having this information facilitated the development of strategies to combat crime (Reyes, 2011). This crowdsourced crime-fighting model has been replicated on a grand scale in the case of the CIC organization (Centro de Integración Ciudadana – Centre for Citizen integration), an organization that gathers a huge amount of information related to criminality from diverse sources like texts, tweets, Facebook posts, and emails. It has now developed a mobile application called "Tehuan," where citizens can make reports, and authorities can gather information about criminality and social problems. As mobile devices became cheaper and more accessible,

Table 7.2 Governmental and communities' use of information technologies

Place	Technology	Use
YouTube Co. (2009)	The use of social media platforms (Facebook, YouTube) to share information about criminality	Use social media to provide information related to organizes crime, as well as to denounce narcocartels or organized crime.
Guerrero State police (2011)	Guerrero. Ciudadanos20 (Citizen Report Management)	A platform to report incidents directly to the police
Ciudad Juárez (2011)	The development of an online map to prevent crimes like TipLine (A US-Mexican project)	An online map that shows high criminality zones, information is provided by citizens
Federal US and Mexican Government (Since 2011)	Drones provided by the US	Surveillance and information gathering to capture narco kingpins.
Jalisco State (2018)	Urban Shield C5	A system of 6,000 cameras. Citizens can share their domestic devices using the internet, so they can be a part of the system
All the country (2019)	Generalization of social media applications: Facebook, YouTube, Twitter, WhatsApp)	Platforms like Facebook Groups or mobile applications like WhatsApp are used to create community groups where people share information: suspects, drug sellers, etc.
Relatives of missing persons (2019)	The use of domestic drones	Drones support the search for human remains (narco fossa) in large and remote areas.

it collected increased information and became an important tool to combat criminality across the country. Thanks to mobile applications like WhatsApp, communities now create and support groups that share information about suspicious people, illicit activities, or urban problems inside their neighborhoods (López Gutiérrez, 2018). There are other cases where citizen groups have taken the initiative to install video camera networks to surveil zones that are dense in criminal activity. Citizens could also connect their local cameras to this network (Reyes, 2019). These actions provide authorities not only evidence to help prosecute criminals but also to introduce preventive actions (See table 7.2).

In recent years, the use of UAV or drones also became more frequent among the police and the army to combat local crime and narcotraffic (Sheridan, 2011). Drones were incorporated to fight organized crime for the same reason that the narco cartels use them: to get a strategic advantage. Drones are ideal devices for surveillance work during the day or night, covering large zones in extreme terrains like jungles, mountains or deserts, or even in crowded urban areas (Konrad, 2017). UAVs allow authorities to monitor the activities of narcocartels at a distance without risking the safety of soldiers. In addition, there are cases of drones used by civilians to search for missing persons, who were kidnapped, killed and disappeared by the criminal organizations. Many times, the criminals buried the bodies of victims in remote rural areas, deterring the recovery of the human remains. However, the use of drones has allowed the relatives of missing people to improve the search for these burial sites, or "Narcofosas" (Balderas, 2019).

Conclusions

This chapter provided a robust narrative on how digital technologies have shaped the drug problem in Mexico. Every actor: the government, the civil organizations, and the narco gangs, have incorporated information technologies to improve the effectivity of their actions. Smart phones and social media allowed the construction of efficient communication channels to gather and share crucial information among the members. Civil organizations, for example, thanks to the WhatsApp's end-to-end encryption technology, were able to coordinate actions with the police forces to combat crime. But at the same time, criminal gangs have taken advantage of these applications to improve the internal communication, limiting the intervention of authorities. A rapid adoption of cutting-age technologies, such as drones and cryptocurrencies, and the inclusion of smartphones and social media, have allowed narco gangs to flourish and globally expand their product far beyond the Mexican borders. They have a strong presence on social medial to gain public sympathy and to set virtual campaigns to discredit the government and other cartels. As it has happened during the Covid-19 pandemic in where criminal organizations distributed food among poor communities and then made public these actions on social media platforms such as Facebook (Barrera, 2020). In addition, the incorporation of drones has become an indispensable element to surveil large rural zones mainly in the northern part of the country, or to find massive graves of missing people in remote geographic areas.

However, as digital technologies evolve and become more accessible and diverse, the processes of technologization of the drug war changes. New devices and systems are incorporated by the different actors to gain substantial advantages. Equipped drones have been used by criminal organizations to attack high-level public force members, and the use of cryptocurrencies has become a popular method among criminal gangs to recover their revenues from the illicit activities. To face this challenging scenery, authorities could eventually incorporate big data analysis and artificial intelligence to surveil and forecast crime (Hao, 2019), or at least to provide a more accurate foresight to develop efficient security policies. In this case, it is urgent to develop well-defined legal frameworks that set limits to the use of emerging technologies as drones or AI platforms to combat crime when this use compromises the peoples' right to privacy and the general protection of human rights. The success or failure of the upcoming security strategies related with the drug war in Mexico will depend on the effective harmonization of the legal framework, major investment in technology, and establishment of effective communication strategies to promote collaboration between the civil organizations and the government.

References

44Lab. (2018). Pugnas internas del CJNG. Esto es lo que sabemos. Retrieved June 2, 2019, from Udgtv website: http://udgtv.com/noticias/pugnas-internas-cjng/

Angelini, D., & Gibson, S. (2007). Organized Crime and Technology. *Journal of Security Education, 2*(4), 65–73. https://doi.org/10.1300/J460v02n04_07

Arsenault, C. (2011). Mexico's narco blog: Drug deaths in real time. *McClatchy - Tribune Business News; Washington.* Retrieved from http://search.proquest.com/abicomplete/docview/863417484/abstract/6BE26C81320643B2PQ/2

Asociación de Internet MX. (2019). *15° Estudio sobre los Hábitos de los Usuarios de Internet en México 2019.* México: Asociación de Internet MX.

Atuesta, L. H. (2017). Narcomessages as a way to analyse the evolution of organised crime in Mexico. *Global Crime, 18*(2), 100–121. https://doi.org/10.1080/17440572.2016.1248556

Axe, D. (2017). Great, Mexican drug cartels now have weaponized drones. *Motherboard.* Retrieved from https://motherboard.vice.com/en_us/article/j5jmb4/mexican-drug-cartels-have-weaponized-drones

Bagley, B. M., & Rosen, J. D. (2015). *Drug Trafficking, Organized Crime, and Violence in the Americas Today.* Gainesville: Univ Pr of Florida.

Balderas, Ó. (2019). Madres de desaparecidos buscan desde el cielo a sus hijos 1/2—MVS Noticias Luis Cárdenas. Retrieved June 30, 2019, from MVS Noticias website: https://mvsnoticias.com/podcasts/segunda-emision-con-luis-cardenas/madres-de-desaparecidos-buscan-desde-el-cielo-a-sus-hijos/

Barrera, J. (2020). Nar-COVID-19. El Informador. https://www.informador.mx/ideas/Nar-COVID-19-20200504-0028.html

Bates, T. (2014). A Mexican drug cartel's rise to dominance. *The Week.* Retrieved from https://theweek.com/articles/452325/mexican-drug-cartels-rise-dominance

BBC. (2008). Guatemala calls for joint action with Mexico to confront drug traffickers. *BBC Monitoring Americas.* Retrieved from https://search.proquest.com/abicomplete/docview/460159136/abstract/E2BCBDDA8ADE46C4PQ/3

BBC. (2019). *Mexico murder rate on the rise again.* Retrieved from https://www.bbc.com/news/world-latin-america-48012923

Bergman, M. (2018). *Illegal Drugs, Drug Trafficking and Violence in Latin America.* Retrieved from//www.springer.com/br/book/9783319731520

Bromby, M. (2006). Security against crime: Technologies for detecting and preventing crime. *International Review of Law, Computers & Technology, 20*(1–2), 1–5. https://doi.org/10.1080/13600860600818235

Bustamante, C. G. de, & Relly, J. E. (2014). Journalism in times of violence: Social media use by US and Mexican journalists working in northern Mexico. *Digital Journalism, 2*(4), 507–523. https://doi.org/10.1080/21670811.2014.882067

Calderón, S. I. (2012). In Mexico, Tech is used to help combat narco violence, Insecurity. Retrieved May 10, 2018, from TechCrunch website: http://social. techcrunch.com/2012/12/25/in-mexico-tech-used-as-aid-to-combat-narco-violence-insecurity/

Castelli, M., Sormani, R., Trujillo, L., & Popovič, A. (2017). Predicting per capita violent crimes in urban areas: An artificial intelligence approach. *Journal of Ambient Intelligence and Humanized Computing*, *8*(1), 29–36. https://doi. org/10.1007/s12652-015-0334-3

Cerda Pérez, P. L., Alvarado Pérez, J. G., & Cerda Pérez, E. (2013). Narco mensajes, inseguridad y violencia: Análisis heurístico sobre la realidad mexicana/Narco Messages, Insecurity and violence: Heuristic Analysis about Mexican Reality. *Historia y Comunicación Social*, *18*, 839–853.

Durán-Martínez, A. (2015). Drugs Around the Corner: Domestic Drug Markets and Violence in Colombia and Mexico. *Latin American Politics and Society*, *57*(03), 122–146. https://doi.org/10.1111/j.1548-2456.2015.00274.x

Emem, M. (2018, December 28). How Mexican Cartels Use Chinese Crypto Brokers to Launder Drug Money. *CCN.Com*. Retrieved from https://www.ccn.com/ how-mexican-cartels-use-chinese-crypto-brokers-to-launder-drug-money/

Engberts, B., & Gillissen, E. (2016). *Policing from Above: Drone Use by the Police*. En B. Custers (Ed.), *The Future of Drone Use: Opportunities and Threats from Ethical and Legal Perspectives* (pp. 93–113). The Hague: Asser Press. https://doi. org/10.1007/978-94-6265-132-6_5

González, A. (2018, May 15). Arma crimen su red telecom. *Reforma*. Retrieved from https://www.reforma.com/aplicacioneslibre/preacceso/articulo/default.aspx-?id=1394355&v=3&urlredirect=https://www.reforma.com/aplicaciones/articulo/ default.aspx?id=1394355&v=3

Green, T. (2014, May 22). Nightmare in the Sky: Drugs via Drones. *Robotics Business Review*. Retrieved from https://www.roboticsbusinessreview.com/unmanned/ nightmare_in_the_sky_drugs_via_drones/

Hallevy, G. (2015). Punishibility of Artificial Intelligence Technology. In *Liability for Crimes Involving Artificial Intelligence Systems* (pp. 185–227). https://doi. org/10.1007/978-3-319-10124-8_6

Hao, K. (2019, February 13). Police across the US are training crime-predicting AIs on falsified data. MIT Technology Review. https://www.technologyreview. com/2019/02/13/137444/predictive-policing-algorithms-ai-crime-dirty-data/

Keene, S. D. (2011). Emerging threats: Financial crime in the virtual world. *Journal of Money Laundering Control*, *15*(1), 25–37. https://doi.org/10.1108/ 13685201211194718

Kenney, M. (2003). From Pablo to Osama: Counter-terrorism Lessons from the War on Drugs. *Survival*, *45*(3), 187–206. https://doi.org/10.1080/00396338.2003.9688585

Konrad, K. D. (2017). HERMES 450: Mexican Air Force Engaged in Fight against Drug Trafficking. *Military Technology*, *41*(7/8), 31–32.

LaSusa, M. (2017, March 27). Mexico's War on Crime: A Decade of (Militarized) Failure. Retrieved June 2, 2019, from InSight Crime website: https://www.insight-crime.org/news/analysis/mexico-drug-war-decade-failure/

Lindo, J. M., & Padilla-Romo, M. (2018). Kingpin approaches to fighting crime and community violence: Evidence from Mexico's drug war. *Journal of Health Economics*, *58*, 253–268. https://doi.org/10.1016/j.jhealeco.2018.02.002

López Gutiérrez, T. (2018, January 20). Así se organizan vecinos para combatir la delincuencia. *Excélsior.* Retrieved from http://www.excelsior.com.mx/nacional/2018/01/20/1214787

Mendoza Rockwell, N. (2010, June 30). El narco y los medios. *Letras Libres.* Retrieved from http://www.letraslibres.com/mexico/libros/el-narco-y-los-medios

Millard, G. H. (1997). Drugs and Organised Crime in Latin America. *Journal of Money Laundering Control, 1*(1), 73–78. https://doi.org/10.1108/eb027122

Monroy-Hernández, A., & Palacios, L. D. (2014). Blog del Narco and the Future of Citizen Journalism. *Georgetown Journal of International Affairs.* Retrieved from https://www.microsoft.com/en-us/research/publication/blog-del-narco-and-the-future-of-citizen-journalism/

Muoio, D. (2016, July 20). Here's all the high-tech gear cartels use to sneak drugs into the US. *Business Insider.* Retrieved from http://www.businessinsider.com/cartels-use-tech-to-sneak-drugs-into-the-us-2016-7

Gavilán, Nicolás (2018). El peligro de ejercer periodismo en México. Análisis de la cobertura informativa del asesinato de Javier Valdez según el enfoque del peace journalism. *The Risks of Practicing Journalism in Mexico. Analysis of the Coverage of the Murder of Javier Valdez According to the Approach of Peace Journalism., 17*(1), 93–113. https://doi.org/10.26441/RC17.1-2018-A5

Nix, J., Smith, M. R., Petrocelli, M., Rojek, J., & Manjarrez, V. M. (2016). The Use of Social Media by Alleged Members of Mexican Cartels and Affiliated Drug Trafficking Organizations. *Journal of Homeland Security and Emergency Management, 13*(3), 395–418. https://doi.org/10.1515/jhsem-2015-0084

Piña-García, C. A., & Ramírez-Ramírez, L. (2019). Exploring crime patterns in Mexico City. *Journal of Big Data, 6*(1), 65. https://doi.org/10.1186/s40537-019-0228-x

Reyes, I. de los. (2011, enero). ¿Puede la tecnología frenar el crimen en México? *BBC Mundo.* Retrieved from http://www.bbc.com/mundo/noticias/2011/01/110125_110125_mexico_crimen_tecnologia_ciudadania_irm

Reyes, H. (2019, March 26). Casi al 80 por ciento el Escudo Urbano C5. Retrieved June 30, 2019, from Quadratin Jalisco website: https://jalisco.quadratin.com.mx/justicia/casi-al-80-por-ciento-el-escudo-urbano-c5/

Schneider, M. (2019, January 29). El Chapo's drug tunnels, explained. *Vox.* Retrieved from https://www.vox.com/videos/2019/1/29/18202097/el-chapo-drug-tunnels-trial-mexico-drug-war

ScienceDaily. (2014, March 4). How social media shaped the "drug war" in Mexico. *ScienceDaily.* Retrieved from https://www.sciencedaily.com/releases/2014/03/140304130027.htm

Selgelid, M. J. (2011). Securitization Technologies. In J. Kleinig, P. Mameli, S. Miller, D. Salane, & A. Schwartz (Eds.), *Security and Privacy* (pp. 89–128). Retrieved from http://www.jstor.org/stable/j.ctt24h8h5.11

Sheridan, M. B. (2011, March 16). Mexico confirms use of U.S. drones in drug war. *Washington Post.* Retrieved from https://www.washingtonpost.com/world/mexico-confirms-seeking-us-drone-help-in-drug-war/2011/03/16/ABbSEZg_story.html

Sullivan, J. P., Bunker, R. J., & Kuhn, D. A. (2018, July 10). Mexican Cartel Tactical Note #38: Armed Drone Targets the Baja California Public Safety Secretary's Residence in Tecate, Mexico. Retrieved October 13, 2019, from Small Wars Journal website: https://smallwarsjournal.com/jrnl/art/mexican-cartel-tactical-note-38-armed-drone-targets-baja-california-public-safety

Tabor, D. (2014, March 25). Radio Tecnico: How The Zetas Cartel Took Over Mexico With Walkie-Talkies. *Popular Science*. Retrieved from https://www.popsci.com/article/technology/radio-tecnico-how-zetas-cartel-took-over-mexico-walkie-talkies

The Economist. (2011, April 14). The drug war hits Central America. *The Economist*. Retrieved from https://www.economist.com/node/18560287

The Economist. (2018, May 5). Mexico's murder rate heads for a new record. *The Economist*. Retrieved from https://www.economist.com/news/americas/21741604-solutions-proposed-main-candidates-president-are-unconvincing-mexicos-murder-rate

Vilalta, C. (2014). How Did Things Get So Bad So Quickly? An Assessment of the Initial Conditions of the War Against Organized Crime in Mexico. *European Journal on Criminal Policy and Research*, *20*(1), 137–161. https://doi.org/10.1007/s10610-013-9218-2

Woody, C. (2019, September 24). The Coast Guard busted another "narco sub" carrying 12,000 pounds of cocaine. *Business Insider*. Retrieved from https://www.businessinsider.com/coast-guard-busts-narco-sub-carrying-12000-pounds-of-cocaine-2019-9

Zepeda Gil, R. (2018). Siete tesis explicativas sobre el aumento de la violencia en México. *Política y Gobierno*, *25*(1), 185–211.

Part III

Labor Markets, digital media, and emerging technologies: Potentials and risks

8 Algorithmic law – a legal framework for Artificial Intelligence in Latin America

Maximiliano Marzetti

Risks and opportunities of Artificial Intelligence in Latin America

The fourth industrial revolution

Schwab (2016), is a new productive paradigm dominated by intelligent autonomous machines characterized by the use of Artificial Intelligence (AI). This disruptive technology can not only substitute human muscle but also human brains. Undeniably, the "gale of creative destruction" (Schumpeter, 1943) is the essence of capitalism. The potential of AI to upset the *status quo*, to create new winners and losers explains both the support and animadversion towards AI.

AI has been defined as "the science of making computers do things that require intelligence when done by humans" (Copeland, 2000). AI may consist of software or a combination of software and hardware, such as self-driving cars, automated weapons systems, and even household robots, like the humble *Roomba*. AI technology has passed through *summers and winters*, i.e. periods of wax and wane, since the notion itself was first introduced by John McCarthy in the 1950s. Currently, the leading AI technologies are *machine learning* and *neural networks*.

The data-driven *algorithmic economy* may bring prosperity or catastrophe to Latin America. A report by Accenture emphasizes that AI could help solve two of the region's endemic problems: the over reliance on the export of commodities in international trade and the productivity deficit, total factor productivity increases in Latin America have been much lower than in other regions (Ovanessoff & Plastino, 2017). Another report authored by an Argentinian think tank concludes that the countries that more rapidly adapted to subsequent phases of previous industrial revolutions grew the most, according to historical data. The same report concludes that early adoption of AI technologies in Argentina could lead to an increase in economic growth of up to 4.4% in the next decade (Abdala, M. B., Eussler, S. L., & Soubie, S., 2019).

Other reports paint a bleaker outlook for the region. For instance, a report by *PricewaterhouseCoop*er suggests Latin America will benefit less than other regions of the world from AI-related initiatives (PricewaterhouseCooper, 2017). Even if due to AI global GDP may increase by 14%, or USD 15.7 trillion by 2030, almost half of this increase will occur in China alone (USD 7 trillion in GDP). North America would be next with an expected increase of 14.5% in GDP (USD 3.7 trillion); followed by Northern European countries (USD 1.8 trillion), Africa, Oceania, and Asian countries excluding China and Japan, (USD 1.2 trillion), developed Asian countries (USD 0.9 trillion), Southern European countries (USD 0.7 trillion) and, at the end, Latin American countries (USD 0.5 trillion). A more recent report on the macroeconomic effects of AI, also by *PricewaterhouseCooper*, maintains the same growth expectations for Latin America (PricewaterhouseCooper, 2018). In addition, a report by *Iniciativa Latinoamericana por los Datos Abiertos* (ILDA, 2018). suggest Latin American governments have failed to implement *open data standards,* which the report identifies as key to promote AI in the region.

However, most reports coincide that the impact of AI in the job market might be considerable. A report by *McKinsey Global Institute* estimates that by 2030 up to 1/3 of current jobs could be lost due to AI (McKinsey Global Institute, 2017). This report also suggests a discrete job creation potential of AI. By 2030, between 75 to 375 million workers will need to switch occupational categories, the report concludes. To date, few Latin American governments have supported AI-related initiatives. A report by the *World Wide Web Foundation* refers to publicly funded AI applications in Argentina, to predict school dropout and teenage pregnancy rates in marginalized neighborhoods in Salta province, and Uruguay, to predict areas where crimes may occur (World Wide Web Foundation, 2018). *Prometea* is another Latin American AI project worth mentioning, to assist the work of the Argentinian judiciary (Corvalán, 2015).

Oppenheimer warns that Latin American countries should take AI seriously and prepare for the socio-economic changes the algorithmic economy may bring about (Oppenheimer, 2018). Latin America is the most unequal continent of the world (World Economic Forum, 2016). Without proper regulation, AI has the potential increase wealth inequality and sub-optimal development (Pasquali, 2019).

The role of government and public policy

Governments provide valuable public goods in the form legal rules. Stable, fair, and efficient legal rules provide *legal certainty* and a stable framework to conduct business (foreseeability). There is consensus in the *Law and Economics* literature that without property rights, enforcement of contracts, and effective Court proceedings no market would function properly (Posner, 1998; Cooter & Schäfer, 2012; Litan et al., 2011). Lately, to foster investments in AI innovations some governments have begun to disclose AI strategies in order to coordinate public and private initiatives, and provide clear signals to market actors. This decision seems to be strategic; governments compete to attract capital and talent. In a global economy companies make decisions such as where to set up new R&D facility taking into account, among other variables, regulation. We may call this phenomenon *regulation shopping*.

Public policies to promote AI may be a subset of a more general *innovation policy*, which has been defined as "public intervention to support the generation and diffusion of innovation, whereby an innovation is a new product, service, process or business model that is to be put to use, commercially or non-commercially" (Edler, J., Cunningham, P., Gök, A., & Shapira, P., 2016). Governments play a leading role in promoting innovation. A report by the International Energy Agency rightly states, "policy affects every link of the innovation chain." Government initiatives may reinforce *technology push* by coordinating publicly funded R&D, avoiding duplication of activities and waste of limited resources. In addition, government policies could reinforce *market pull*, for instance by providing tax incentives to startups and entrepreneurs.

AI policy must be designed for the wellbeing of human beings, what I call *human-centric AI policymaking*. In 2019 the *Organization for Economic Co-operation and Development* (OECD) adopted its *Principles on Artificial Intelligence*. The same institution later provided a set of principles to guide AI policy, this are: human-centeredness, promotion of inclusive and sustainable growth, algorithmic transparency and explainability, robustness, security, safety, accountability, *inter al.* (OECD Publishing, 2019). Of course, effective policymaking requires an important dose of pragmatism. Drafting an AI policy per se would not suffice to bring about investments, jobs, and growth. Brynjolfsson and McAfee rightly suggests that to promote AI governments should take more general actions, including other basal areas, such as improving the education system, supporting startups, investing in infrastructure, and redesigning the tax system, *inter al.* (Brynjolfsson & McAfee, 2014).

Canada, China, Denmark, the EU, Finland, France, India, Italy, Japan, Mexico, the Nordic-Baltic region, Singapore, South Korea, Sweden, Taiwan, the United Arab Emirates, and the United Kingdom have been the first countries to provide *AI framework policies or national strategies.* (Lauterbach, 2019*).* According to a report by *Oxford Insights* (Oxford Insights, 2019). on the state of governmental AI readiness, Latin American countries are lagging. Only Mexico (ranked no. 32) and Uruguay (ranked no. 35) are mentioned in the report. Mexico disclosed the *Estrategia de Inteligencia Artificial MX in 2018 (*Zapata, 2018, March 27). and Uruguay published the *Estrategia de Inteligencia Artificial para el Gobierno Digital* in 2019 (Presidencia Republica Oriental de Urguay, n.d.). Perhaps other Latin American countries will follow suit.

Algorithmic law

In this article, I use the term *algorithmic law* to refers to principles and legal rules intended to maximize the opportunities derived from AI while minimizing its associated risks and potential negative externalities. Consequently, some of these legal rules are *enabling*, i.e. promoting investments in AI innovations, while others are *limiting*, in an attempt to reduce the potential negative side effects of disruptive innovations such as their impact on human rights and labor relations, to name a few.

Accordingly, *algorithmic law* is not a new area of law but its transversal to many, it encompasses different legal rules across legal branches, from criminal and civil liability, to intellectual property rights, and competition law. In the following paragraphs I will review some of these areas. However, my aim is not to be thorough, but to highlight areas of the law that may require the immediate attention of the Latin American regulator.

Patent law

Patent law incentivizes investments in innovations. *Ex ante*, the decision to invest in an innovation is risky and commercial success uncertain. Patents represent a regulatory solution to the market failure of public goods (valuable information is non-rival and non-excludable). Moreover, Cooter suggests that patents solve the *double trust dilemma*: in the absence of a legal entitlement an inventor has no incentive to disclose his invention; without disclosure no bank or venture capitalist would be willing to invest in it (Cooter, 2014). Thus, patent facilitates investments in innovations.

A WIPO Report shows AI-related inventions have been on the rise since 2013 (WIPO, 2019). The ratio of scientific papers to inventions in the area of AI fell from 8:1 in 2010 to 3:1 in 2016. This fact, coupled with the increasing number of patent applications in the area, confirms a shift from basic science to scientific applications, which is signal of a mature technology. Machine learning-related patent applications have grown by an average of 26% annually between 2011 and 2016 and represent almost one third of all AI-related patent applications (the report surveyed a total of 134,777 patent documents). Companies from the US, Japan, and China dominate worldwide patent activity, with IBM and Microsoft being the two companies with the largest AI-related patent portfolios. The sectors with the most AI-related patents are telecommunications (15%), transportation (15%), life and medical (12%), and human-computer interaction (11%).

Because patents are a legal monopoly, patentability requirements are strict, *viz.*, absolute novelty, inventive step (non-obviousness), and industrial applicability. These requirements, fit for the first industrial revolution, may not be well suited for the fourth one. The WIPO report also describes some challenges AI-related innovations may face to obtain a patent (WIPO Standing Committee on the Law of Patents, 2019). Firstly, novelty is difficult to assess given the limited amount of available prior art (AI-related innovations remain at an early stage of development). Secondly, the determination of the inventive step requirement requires a person having ordinary skill in the art, a legal standard difficult to meet in relation AI technologies. Thirdly, the extent of enabling disclosure in a patent application is unclear. For instance, in an AI-related patent application, do algorithms, training models, datasets, neural network architectures, etc., have to be disclosed too? To incentivize AI-related inventions, patent laws and patent examination guidelines may have to be revised. However, the regulator must be cautious. If market incentives and other strategies to appropriate value are sufficient, granting patents to AI tools should be denied, however, they may be necessary to protect/incentivize AI outputs (Hilty, Hoffmann & Scheuerer, 2020).

Another area that calls for attention is whether AI systems can be considered inventors (and even authors, see *infra*). In general terms, most national laws bar AI systems from being considered inventors and/or patent holders. On April 22, 2020 the US Trademark and Patent Office rejected a patent application (serial no. 16/524,350 filed July 29, 2019) in which the denominated inventor was an AI system (identified as *Dabus*). Also, and in the same direction, in 2020 the European Patent Office rejected two patent applications designating an AI system as inventor (EP 18 275 163 and EP 18 275 174). The only reasonable rationale to grant inventor or owner status to a machine, from a human-centric perspective, would be to increase incentives to investments in AI R&D. To that end, most studies suggest inventorship status and/or ownership of the resulting patent should be given to the person owning or controlling the AI.

Latin America does not take part in the global race for patenting AI-related inventions. Actually, patent applications by Latin American residents are extremely low in all areas of technology. Adding up all the patent applications of 20 Latin American countries filed in 2016 they represent 2.2% of all the patent filings worldwide for the same year (Llorens, 2017). Moreover, the majority of the patent applications filed in the region are by non-residents. According to statistics of the World Bank, in 2017, residents applied for 393 patents in Argentina (of a total of 3050), 425 in Chile (of a total of 2469), 595 in Colombia (of a total of 1777), 5480 in Brazil (of a total of 20178), 1334 in Mexico (of a total of 15859), and 100 in Peru (of a total of 1119). For perspective, patent applications by residents for the same year in China were 1.245.709, in South Korea 159.084, in Japan 260.290, and in the United States 293.904.

Although it is possible to innovate without patents, it is very difficult to recover R&D investments or secure a sustainable competitive advantage without them. Moreover, for startups, patents are a determinant factor to obtain seed capital (Cao, Hsu, Bengtsson, et al. 2010). Certainly, Latin American governments should do more to incentivize their nationals to file for patents, domestically and abroad, in AI and other sectors of technology.

Copyright law

Copyright protects creative works (texts, songs, films, and even software). Copyright protection last longer than a patent, usually the life of the author plus 50 to 100 years *postmortem auctoris*, but its scope of protection is more limited as only original expressions are protected but not ideas. Like patents, copyright law can incentivize investments in AI R&D, provide legal certainty to companies and signals to investors.

AI systems have already created original works in a somewhat autonomous way, such as paintings mimicking the style of Rembrandt, new musical piano compositions or even novels in Japanese (Guadamuz, 2017). Should we consider AI systems authors, from a copyright law perspective? Like in the case of patents, jurisdictions answer this question in the negative. A recent case involving a crested macaque, named Naruto, who allegedly took pictures of himself (i.e. selfies) using a field camera may provide, *mutatis mutandis*, some guidance. After protracted litigation, the *US Court of Appeal for the Ninth Circuit* decided that even if the macaque had constitutional standing to be in Court, represented by the association People for the Ethical Treatment of Animals, it had not statutory standing to claim copyright infringement (Naruto v. Slater, No. 16-15469, 9th Cir., 2018). After this decision the *US Copyright Office* updated its rules of authorship:

> [t]he US Copyright Office will register an original work of authorship, provided that the work was created by a human being. The copyright law only protects 'the fruits of intellectual labor' that 'are founded in the creative powers of the mind.' ... Because copyright law is limited to 'original intellectual conceptions of the author,' the Office will refuse to register a claim if it determines that a human being did not create the work (US Copyright Office, 2014).

The decision in Naruto and the updated guidelines of the *US Copyright Office* seem to suggest AI systems will not be deemed authors either. However, AI-generated works may be considered *works for hire*, in which case the owner of the AI system is considered the author. This doctrine, coherent with the goal to incentivize investments in AI, is even applied in civil law countries in relation to software created under an employment relationship. For instance, article 4 d of Argentinian Copyright Act no. 11.723, modified by Act no. 25.036. The *UK Copyright, Designs and Patents Act 1988* includes an interesting provision in Section 12. In the case of a computer-generated literary, dramatic, musical, or artistic work, the author shall be taken to be the person by whom the arrangements necessary for the creation of the work are undertaken, in which case protection extends for 50 years since the work was created and no moral rights are recognized.

However, copyright law plays a fundamental role to the development and education of AI algorithms. Neither humans nor machines create *ex nihilo*. Raw data is rarely sufficient, in most circumstances, to train AI algorithms. Like a human child, an AI system needs to access literary, scientific and artistic works to learn to think. Most recent works are copyrighted, which makes access prohibitively expensive due to high transaction (Coase, 1960) and financial costs. In the absence of a specific *ex lege* permission, called exceptions or limitations to copyright, unlicensed reproduction of copyrighted contents is unlawful and may lead to a copyright infringement lawsuit. Existing exceptions and limitations to copyright law have been thought with a human user in mind and are not applicable to AI systems.

Thus, specific exceptions and limitations may be necessary to facilitate training and education of AI systems. One AI technology in particular, text and data mining, may require access to large quantities of copyrighted works. To remedy this situation, some countries have already amended their copyright regimes. In 2014 the UK amended the *Copyright, Designs and Patent Act of 1988* to add an exception to allow computational analysis of anything recorded in a copyrighted work, of which the user has lawful access, for the sole purpose of research for non-commercial purposes (Section 29A of the UK Copyright, Designs, and Patent Act 1988). In 2018 Japan also amended its copyright law to support investments in AI systems (Japanese Copyright Act no. 48 of May 6, 1970, amended by Act No. 72 of July 13, 2018). Current article 30–4 of the *Japanese Copyright Act* allows access to copyrighted works for data analysis (including extraction, comparison, classification, or other statistical analysis of language, sound, or image data, or other elements of which a large number of works or a large volume of data is composed) and computer data processing; article 47–-4 allows the making of electronic incidental copies of works; and article 47–-5 permits the use of copyrighted works for data verification purposes in connection to research.

The US *fair use doctrine*, incorporated into Section 107 of the *US Copyright Act*, is more flexible than the system of exceptions and limitations of civil law countries. Fair use gives judges four parameters to determine, in a concrete case, whether a certain unauthorized act infringes copyright or not. The US *fair use* has been effectively used to upheld new technologies that had potential infringing uses, such as the recording copyrighted works using videocassette recorders. In the famous *Betamax* case decided in 1984, the US Supreme Court stated that recording copyrighted works for time-shifting purposes was compatible with fair use (Sony Corp. of America v. Universal City Studios, Inc., 464 US 417, 1984). In 2015 the Court of Appeals of the Second Circuit found that *Google Books'* limited digitization of copyrighted works was also compatible with fair use (Authors Guild v. Google, Inc., 804 F.3d 202, 2015). Some authors suggest the *Google Books* precedent could be used to justify reproducing large quantities of copyrighted contents to train AI systems, while others are skeptical (Sobel, 2017).

In the EU, a number of directives were passed to approximate copyright and related rights of member countries. Some of these directives have included mandatory and voluntary exceptions and limitations. It is worth noting that the only mandatory exceptions for members countries to implement are those related to technology. *Directive 2001/29/EC of the European*

Parliament and of the Council of 22 May 2001 on the harmonization of certain aspects of copyright and related rights in the information society incorporated in art. 5.1 an exception to protect Internet intermediaries from infringement for making transient reproductions of copyrighted contents.

> Art. 5 1. Temporary acts of reproduction referred to in Article 2, which are transient or incidental [and] an integral and essential part of a technological process and whose sole purpose is to enable: (a) a transmission in a network between third parties by an intermediary, or (b) a lawful use of a work or other subject-matter to be made, and which have no independent economic significance, shall be exempted from the reproduction right provided for in Article 2.

More recently, *EU Directive 2019/790 of the European Parliament and of the Council of 17 April 2019 on copyright and related rights in the Digital Single Market* introduced two mandatory exceptions, one for text and data mining (TDM) for scientific research (article 3) and other purposes (article 4).

TDM mining refers to computer-based analysis of large bodies of data to extract meaningful patterns and train AI algorithms. Hugenholtz suggests that "much of the current and future development in artificial intelligence, therefore, depends on text and data mining" (Hugenholtz, 2019). The new TDM exceptions permit the reproduction and extraction of lawfully accessible copyrighted contents, necessary to train AI systems, without requiring previous authorization or the payment of royalties to the copyright holder. Because these exceptions refer to non-expressive uses of copyrighted works, they do not harm the copyright holder. TDM exceptions may also serve to promote other goals, such as to avoid the creation of biased algorithms that may lead to unlawful discrimination. For instance, an AI system deprived from access to a sufficiently broad pool of resources, may produce biased output (Levendowski, 2018). However, some scholars claim the new EU mandatory exceptions are insufficient. Firstly, they may be too narrow, as they only refer to acts of reproduction, leaving outside of its scope acts of communication or making available to the public. Secondly, copyright laws are national whereas investments in R&D are becoming increasingly transnational. For these reasons, a group of academics has suggested it may be time to discuss a multilateral treaty for AI, under the auspices of the World Intellectual Property Organization (Flynn et al., 2020).

The situation of exceptions and limitations to copyright law in Latin America is asymmetric. For instance, the Argentinian Copyright Act no. 11.723, enacted in 1933, is more restrictive. It contains no explicit exception or limitation for technological processes, neither for libraries nor archives, which is an oddity in the comparative landscape (Crews, 2017). In contrast, the Chilean (article 71 Chilean Copyright Act. no. 17.336) and Brazilian (articles 46, 47, and 48 of the Brazilian Copyright Act no. 9.610) laws have been modified recently to enlarge the list of exceptions and limitations. In any case, no Latin American country, has incorporated specific exception or limitations for AI systems. Failing to adapt copyright law to new technologies and social mores places commonplace actions of the majority of citizens in a dangerous zone of illegality or *a-legality*.

Last but not least, Argentina, Paraguay, and Uruguay still enforce a *domaine public payant*, a system that requires the payment of a tax to a state agency to exploit works in the public domain, this may become an additional hurdle for the training of AI systems (Marzetti, 2019).

Trade secrets

Industrial espionage and trade secret theft is a threat to the competitiveness of a private firms and entire economies. The Agreement on Trade-Related Aspects of Intellectual Property Rights (TRIPS) obliges members to the World Trade Organization to protect undisclosed information, as long as it is secret, has commercial value, and has been subject to reasonable measures to keep it secret (art. 39). Sappa suggests effective trade secret protection may promote investments in AI (Sappa, 2019).

This seem to be the reason behind the recent trend of reinforcing protection for trade secrets. In 2016 the US enacted the *Defend Trade Secrets Act* and the EU passed *Directive 2016/943 on the protection of undisclosed know-how and business information (trade secrets) against their unlawful acquisition, use and disclosure*. And one year later, China amended its *Unfair Competition Law* to reinforce trade secret protection. Conversely, in Latin American countries rules on trade secret protection have not been revised after the TRIPS Agreement was ratified: Argentina's Act no. 24.766 on Confidential Information dates back to December 18, 1996 and Brazil's Law Industrial Property no. 9.279, which protects trade secrets by way of unfair competition provisions, dates back to May 14, 1996 (amended in 2001).

Personal data

In most countries the protection of privacy has become a fundamental right. Personal data is s any information that relates to an identified or identifiable human person. The EU *General Data Protection Regulation 2016/679* (GDPR) is the most comprehensive data protection law in the world. To what extent strong protection of personal data is favorable or detrimental to the development of AI systems? Most AI systems improve their accuracy exponentially the larger the amount of data they process, which allows them to establish more robust and unanticipated correlations. However, the GDPR provides for a very stringent basis for the lawful processing data. Consent, for instance, has to be freely given for one or more specific purpose (art. 7). Moreover, consent can be also freely withdrawn, and, under certain circumstances, the data subject may request her personal data to be erased (art.17). The more rights are granted to human persons, the more costly compliance becomes for companies. The *Center for Data Innovation* stated in a 2018 report that the GDPR will put the EU at a competitive disadvantage *vis-à-vis* the US or China in the race for AI-dominance (Wallace & Castro, 2018).

In the US there is no comprehensive data protection law. At the federal level, scattered rules exist, such as the *Video Privacy Protection Act* of 1988, the *Cable Television Protection and Competition Act* of 1992, the *Fair Credit Reporting Act* or the *Children's Online Privacy Protection A*ct of 1998, to name a few. This approach may prove more favorable to businesses, and less protective of data subject's rights. Japan has also its own model for the protection of personal information. Even if data protection laws exist in Japan, Ishii suggests Japanese companies prefer soft rules, like the concept of *privacy by design*, in order to deal with transparency, consent, prevent black box systems, algorithmic opacity, discriminatory automatic profiling, etc. (Ishii, 2019).

Most Latin American countries have enacted their own privacy laws. The Argentinian data protection Act no. 25.326 dates back to 2000. Conversely, in 2018 Brazil enacted a new General Data Protection Act (LGPD), which will enter into force in August 2020. This new Act replaces previous federal rules and is intended to reinforce the rights of data subjects but at the same time to bring legal certainty to companies. The LGPD is inspired by the GDPR, but the LGPD may be more favorable to companies. For instance, the GDPR imposes a strict 72h term to report data breaches, while the LGPD has no such deadline. Also, in case of non-compliance, fines are significantly lower in the LGPD than in the GDPR. Finally, the LGPD incorporates ten lawful bases for processing personal data, while the GDPR establishes only six.

Competition law

Antitrust scholars are concerned of AI's potential to infringe competition law, algorithms may favor collusion and make it difficult prove (Petit, 2017; Deng, 2018; Surblyte, 2016). Companies such as Google, Amazon, Facebook, and Apple, having access to a huge pool of consumer information, may make perfect price discrimination possible, with adverse effect on consumer welfare. New entrants may not be able to compete in data-driven markets unless they can access sufficient data pools (Kerber, 2018). At the moment it is unclear whether the essential facility doctrine may be applicable to data (Massadeh, 2011).

Proposals to maintain data-driven markets competitive and contestable range from the creation of new forms of IP right for non-personal data (Kerber, 2016), to more aggressive uses of existing competition law tools to force companies to share data with competitors (Drexl, 2017). The Japanese government suggested to create a *right of data portability* to increase competition in the data-driven economy (Hayashi & Arai, 2019), solution similar that found in art. 20 of the GDPR.

Latin American countries have a long tradition of closed economies and weak enforcement of competition law. Fortunately, this trend started to change in the last decade. Governments should update their competition policies in preparation for the algorithmic economy.

Employment law

Will AI bring the end of human labor? Probably not, at least not in the short and medium term. Nevertheless, jobs will be at risk. How well prepared are Latin American countries to face the *new social question*? The risk of job destruction by AI seems to be warranted. A 2014 study by a European think tank estimated that 54% of jobs in the Eurozone are threatened by AI and automation (Bowles, 2014, July 24). The most pessimistic view suggest AI will lead to mass unemployment and more wealth concentration (Ford, 2016). Developing countries may be more vulnerable than developed ones due to the preponderance of low skilled jobs. For instance, while self-driving trucks already exist, the largest union in Argentina remains that of truck drivers (Sindicato de camioneros, n.d). Benefits for companies are evident, self-driving trucks get no salary, take no holidays and will never strike. In the *fourth industrial revolution* white collar jobs are threatened too. Susskind suggests the legal as well as other professions will radically change as a consequence of AI (Susskind, 2008; 2013; Susskind & Susskind, 2015). Low-cognitive quality, repetitive and commodifiable legal work is already being taken over by AI, paralegals and junior lawyers are the most threatened legal professionals at the moment. However, there is potential for job creation too, especially for tasks that require human skills (Frey & Osborne, 2017).

A report published by the *International Labor Organization* shows moderate optimism about the future of employment, assuming stakeholders invest in re-educating workers and governments update their legislations accordingly (ILO, 2018). Some Latin American unions are already preparing themselves for the algorithmic economy. For instance, the *Argentinian Federation of Employees of Commerce and Services* has created the *Artificial Intelligence Institute for the New Argentinian Development* (FAECYS, 2019). This *Institute* has published a report titled *Technological and Algorithmic Social Justice*. The report is clearly defensive and calls for redistributive measures, it even seems to suggest passing to bar the implementation of AI in certain sectors: "Big Data, AI, the IoT, among many other technologies, should be viewed from the perspective of fundamental values, from which important ethical dimensions must be considered, including the primary decision to use or not use these technologies in certain sectors." The translation belongs to the author. This proposal, if put into practice, may seriously affect the competitiveness of companies.

Automated hiring systems is an AI-based implementation that, without the proper legal framework, may result in discriminatory hiring practices. Since most selection algorithms are opaque and not disclosed these practices are difficult to detect. The GDPR offers a palliative against *automated hiring systems*, providing the "data subject shall have the right not to be subject to a decision based solely on automated processing, including profiling, which produces legal effects concerning him or her or similarly significantly affects him or her" (art. 22). However, human oversight may not be sufficient. Thus, some scholars suggest incorporating a *right to an explanation*, to check whether the algorithmic decision respects existing legal standards (Sánchez-Monedero, Dencik & Edwards, 2019).

In this stage of uncertainty, it is useful to turn to history. Gutenberg's movable type printing press, invented in Germany *circa* 1450, was a disruptive technology that led to a tremendous boost of book productivity. Before it, books took months to produce, being patiently copied by hand. The press made possible the reproduction of entire books in only a few hours and as a consequence books became cheaper. This technological breakthrough put the amanuensis out of job. However, this did not happen from one day to another, according to Peter Yu (Yu, 2006). This author suggests amanuenses, scribes, and book illuminators coexisted with the printing press for many decades. Yu estimates that amanuenses coexisted with the press for at least a century. This was probably due to the lifecycle of technology (early presses were likely costlier than human labor), which may have allowed some amanuenses to acquire new skills and successfully transition to new occupations, such as secretaries, stenographers, notaries, etc. This is a valuable historical lesson government should take notice of. Current employment and labor laws, based in the Fordist models of labor, are ill equipped to deal with contemporary challenges, such as the gig economy or AI (Hendrickx, 2018). Governments should begin to design policies educate new generations and retrain workers in order to give them the skills they will need to work in the algorithmic economy.

Civil and criminal liability

Are existing extracontractual liability rules (torts) applicable to harm caused by AI-systems? Intentional standards, such as *dolus*, formulated with a human actor in mind, seem currently inapplicable to intelligent machines. However, will machines, at some point, became capable of autonomous acts of volition? Strict liability or no-fault standards seem less problematic and engage the responsibility of the owner or person responsible the AI system. For instance, art. 1757 of the Argentinian Civil and Commercial Code, which states, "[e]very person is responsible for the damage caused by the risk or vice of things, or activities that are risky or dangerous due to their nature, the means used or the circumstances of their performance," might be applicable, *mutatis mutandis*, to harm caused by AI systems.

In a famous paper Calabresi suggested a goal of torts law is to minimize the cost of accidents in society (Calabresi, 1977). Sanctions may have a deterrent effect on rational human beings, but what about intelligent machines?

The *European Civil Law Rules on Robotics* (Nevejans, 2016) suggest an ethical approach (*soft law*) instead of hard legal rules, in its *Charter on Robotics*. Some countries are evaluating passing specific provisions to address the risks posed by autonomous vehicles (Keller, 2018). The states of Nevada, California, and Florida have already authorized self-driving cars, under certain conditions, to circulate in public roads. What about the legal status of AI? Atabekov and Yastrebov claim the Saudi government has granted the Saudi citizenship to a robot called *Sofia* and Japan gave a residence permit to the chatbot *Shibuya Mirai* (Atabekov, & Yastrebov, 2018).

Criminal law has held for long time that legal persons could not be criminally liable, i.e. *societas delinquere non potest*. However, the situation has changed in recent years. Today the legislations of the US, France, and the UK, to name a few, state that legal persons can be held criminally liable and subject to criminal sanctions, which range from fines to dissolution (equivalent to the death sentence of a human person). Would killer robots may be held criminally liable in the future? Some legal scholars have already developed an analytical framework to justify imposing criminal liability and punishment to machines, including permanent incapacitation for serious offenses, under the doctrine of *machina sapiens criminalis* (Hallevy, 2013).

These and other queries will require serious attention by Latin American policymakers.

Concluding remarks

Latin American governments must prepare for the imminent impact AI in their economies and societies. AI poses risks but also opportunities. Latin America remains the most unequal continent of the world. Without the proper legal framework, AI may increase this inequality. Nonetheless, AI has also the potential to ramp up productivity and enhance the competitiveness of Latin American firms in the global marketplace. To avoid threats and seize opportunities, governmental leadership and a proper regulatory framework are fundamental. I call *algorithmic law* the set of legal rules intended to maximize the advantages derived from AI, while minimizing its disadvantages. The areas and legal problems discussed in this article are only a sample of them. The moment to seriously discuss AI policies has come, the *fourth industrial revolution* is just around the corner.

References

Abdala, M. B., Eussler, S. L., & Soubie, S. (2019). *La política de la Inteligencia Artificial: Sus usos en el sector público y sus implicancias regulatorias* (p. 25). CIPPEC.

Atabekov, A., & Yastrebov, O. (2018). Legal Status of Artificial Intelligence Across Countries: Legislation on the Move. *European Research Studies Journal*, 4, 773–782.

Bowles, J. (2014). Chart of the Week: 54% of EU jobs at risk of computerization [Blog]. https://www.bruegel.org/2014/07/chart-of-the-week-54-of-eu-jobs-at-risk-of-computerisation/

Brynjolfsson, E., & McAfee, A. (2014). *The Second Machine Age: Work, Progress, and Prosperity in a Time of Brilliant Technologies*. W. W. Norton.

Calabresi, G. (1970). *The Costs of Accidents: A Legal and Economic Analysis*. New Haven: Yale University Press.

Cao, J. X., Hsu, P. H., Bengtsson, O., et al. (2010). *Patent Signaling, Entrepreneurial Performance, and Venture Capital Financing*.

Coase, R. H. (1960). The Problem of Social Cost. *The Journal of Law and Economics*, 3, 44.

Cooter, R. (2014). *The Falcon's Gyre: Legal Foundations of Economic Innovation and Growth*. Berkeley Law Books.

Cooter, R. & Schäfer, H.-B. (2012). *Solomon's Knot: How Law Can End the Poverty of Nations*. Princeton University Press;

Copeland, J. (2000). *What is Artificial Intelligence?* Retrieved December 7, 2019, from http://www.alanturing.net/turing_archive/pages/Reference Articles/What is AI.html.

Corvalán, G. (2015). Artificial Intelligence, Threats, Challenges and Opportunities - Prometea, the First Predictive Artificial Intelligence at the Service of Justice is Argentinian. *Revue Internationale de Droit Des Données et Du Numérique*, 3(0), 69–78.

Crews, K. D. (2017). *Study on Copyright Limitations and Exceptions for Libraries and Archives: Updated and Revised*.

Deng, A. (2018). *When Machines Learn to Collude: Lessons from a Recent Research Study on Artificial Intelligence*.

Drexl, J. (2017). *Designing Competitive Markets for Industrial Data Between Propertisation and Access*. JIPITEC.

Edler, J., Cunningham, P., Gök, A., & Shapira, P. (Eds.). (2016). *Handbook of innovation policy impact*. Edward Elgar.

Federación Argentina de Empleados de Comercio y Servicios (FAECYS). (2019). *Conocimiento estratégico para el futuro del trabajo mercantil*.

Flynn, S., Geiger, C., Quintais, J., Margoni, T., Sag, M., Guibault, L., & Carroll, M. W. (2020, April 20). *Implementing User Rights for Research in the Field of Artificial Intelligence: A Call for International Action*.

Ford, M. (2016). *The Rise of the Robots: Technology and the Threat of Mass Unemployment*. New York: Oneworld Publications.

Frey, C. B., & Osborne, M. A. (2017). The future of employment: How susceptible are jobs to computerisation?. *Technological forecasting and social change*, 114, 254–280.

Guadamuz, A. (2017). Artificial intelligence and copyright. *WIPO Magazine*, vol. 5, pp. 14–19

Hallevy, G. (2013). *When Robots Kill: Artificial Intelligence Under Criminal Law*. Northeastern University Press.

Hayashi, S., & Arai, K. (2019). How Competition Law Should React in the Age of Big Data and Artificial Intelligence. *The Antitrust Bulletin*, 64(3), 447–456.

Hendrickx, F. (2018). *From Digits to Robots: The Privacy-Autonomy Nexus in New Labor Law Machinery. Comparative Labor Law & Policy Journal*, 40.

Hilty, R; Hoffmann, J. & Scheuerer, S. (2020). *Intellectual Property Justification for Artificial Intelligence*, Max Planck Institute for Innovation & Competition Research Paper No. 20–02.

Hugenholtz, B. (2019). *The New Copyright Directive: Text and Data Mining (Articles 3 and 4)* - Kluwer Copyright Blog. Retrieved April 28, 2020, from http://copyrightblog.kluweriplaw.com/2019/07/24/the-new-copyright-directive-text-and-data-mining-articles-3-and-4/

Iniciativa Latinoamericana por los Datos Abiertos (ILDA). (2018). *Automatizar con cautela. Datos e Inteligencia Artificial en América Latina.*

International Labor Organization (ILO). (2018). *The economics of artificial intelligence: Implications for the future of work.*

Ishii, K. (2019). Comparative legal study on privacy and personal data protection for robots equipped with artificial intelligence: looking at functional and technological aspects. *AI and Society*, 34(3), 509–533.

Keller, P. (2018). *Autonomous Vehicles, Artificial Intelligence, and the* Law. The Journal of Robotics, *Artificial Intelligence & Law*, 1(2).

Kerber, W. (2016). A New (Intellectual) Property Right for Non-Personal Data? An Economic Analysis. *Gewerblicher Rechtsschutz Und Urheberrecht, Internationaler Teil (GRUR Int)*, (11), 989–999.

Kerber, W. (2018). Data Governance in Connected Cars: The Problem of Access to In-Vehicle Data. *Journal of Intellectual Property, Information Technology and Electronic Commerce Law*, 9.

Lauterbach, A. (2019). *Artificial intelligence and policy: quo vadis?* Digital Policy, *Regulation and Governance*, 21(3), 238–263.

Levendowski, A. (2018). How Copyright Law Can Fix Artificial Intelligence's Implicit Bias Problem. *Wash. L. Rev.*, 93(579).

Litan, R. E., Benkler, Y., Butler, H. N., Clippinger, J. H., Cook-Deegan, R., Cooter, R. D., … Wittes, B. (2011). *Rules for Growth: Promoting Innovation and Growth Through Legal Reform.*

Llorens, G. (2017). *R&D and Innovation, A Patent Failure In Latin America.* Retrieved December 7, 2019, from https://www.worldcrunch.com/business-finance/rd-and-innovation-a-patent-failure-in-latin-america.

Marzetti, M. (2019). *Paying for works in the public domain? The "domaine public payant" in the 21st century.* GRUR Int.

Massadeh, A. A. (2011). *The Essential Facilities Doctrine Under Scrutiny: EU and US Perspective* (No. 2011- AM-1).

McKinsey Global Institute. (2017). *Jobs lost, jobs gained: Workforce transitions in a time of automation.* McKinsey & Company. https://www.mckinsey.com/featured-insights/future-of-work/jobs-lost-jobs-gained-what-the-future-of-work-will-mean-for-jobs-skills-and-wages#

Nevejans, N. (2016). *European civil law rules in robotics.*

Ovanessoff, A., & Plastino. (2017) E., Accenture, *How Artificial Intelligence can drive South America's growth.*

Organización para la Cooperación y el Desarrollo Económicos (OECD Publishing). (2019). *Artificial Intelligence in Society.*

Oppenheimer, A. (2018). ¡*Sálvese quien pueda!: El futuro del trabajo en la era de la automatización.* Penguin Random House Grupo Editorial México.

Oxford Insights. (2019). Government Artificial Intelligence Readiness Index. Retrieved from https://www.oxfordinsights.com/ai-readiness2019.

Pasquali, M. (2019, March 11). *Latin America: Gini coefficient income distribution inequality, by country.* Retrieved April 30, 2020, from https://www.statista.com/statistics/980285/income-distribution-gini-coefficient-latin-america-caribbean-country/

Petit, N. (2017). A*ntitrust and Artificial Intelligence: A Research Agenda. Journal of European Competition Law & Practice,* 8(6).

Posner, R. (1998). *Creating a legal framework for economic development. The World Bank Research Observer,* 13(1), 1–11;

Presidencia Republica Oriental de Urguay. (n.d.). Estrategia de Inteligencia Artificial para el Gobierno Digital (p. 16). Agesic Desarrollo Uruguay Digital.

Pricewaterhouse Cooper. (2017). *Sizing the prize. PPwC's Global Artificial Intelligence Study: Exploiting the AI Revolution. What's the real value of AI for your business and how can you capitalize?*

Pricewaterhouse Coopers. (2018). *The macroeconomic impact of artificial intelligence.*

Sánchez-Monedero, J., Dencik, L., & Edwards, L. (2019). *What Does It Mean to 'Solve' the Problem of Discrimination in Hiring?*

Sappa, C. (2019). How data protection fits with the algorithmic society via two intellectual property rights – a comparative analysis. *Journal of Intellectual Property Law & Practice,* 14(5), 407–418.

Schumpeter, J. (1943). *Capitalism, Socialism and Democracy.* Harper and Row.

Schwab, K. (2016). *La cuarta revolución industrial.* Debate.

Sindicato de camioneros. (n.d.). Sindicato de choferes de Camiones. Retrieved June 29, 2020, from https://www.camioneros-ba.org.ar/

Sobel, B. (2017). *Artificial Intelligence's Fair Use Crisis,* 41 Columbia Journal of Law & the Arts 45.

Surblyte, G. (2016). Data-Driven Economy and Artificial Intelligence: Emerging Competition Law Issues. *Wirtschaft Und Wettbewerb,* 67(3), 120–127.

Susskind, R. (2008). *The End of Lawyers? Rethinking the Nature of Legal Services.* Oxford University Press.

Susskind, R. (2013); *Tomorrow's Lawyers: An Introduction to Your Future.* Oxford University Press

Susskind, R. E., & Susskind, D. (2015). *The Future of the Professions: How Technology Will Transform the Work of Human Experts.* Oxford University Press.

US Copyright Office. (2014). *Compendium of US Copyright Office Practices,* Third Edition, Section 306 ("The Human Authorship Requirement").

Wallace, N., & Castro, D. (2018). *The Impact of the EU's New Data Protection Regulation on AI.*

WIPO. (2019). *Technology Trends – Artificial Intelligence.*

WIPO Standing Committee on the Law of Patents. (2019). *Background document on patents an demerging technologies* (Thirtieth Session Geneva, June 24 to 27, 2019).

World Economic Forum. (2016). *Latin America is the world's most unequal region. Here's how to fix it* .Retrieved December 6, 2019, from https://www.weforum.org/agenda/2016/01/inequality-is-getting-worse-in-latin-america-here-s-how-to-fix-it/.

World Wide Web Foundation (WWWF). (2018). *Algoritmos e Inteligencia Artificial en Latinoamérica - Un Estudio de implementaciones por parte de Gobiernos en Argentina y Uruguay.*

Yu, P. K. (2006). Of Monks, Medieval Scribes, and Middlemen. *Mich. St. L. Rev.,* 1.

Zapata, E. (2018, March 27). Estrategia de Inteligencia Artificial MX 2018. Datos Abiertos MX. https://datos.gob.mx/blog/estrategia-de-inteligencia-artificial-mx-2018?category=noticias&tag=nula.

9 Automation and robotization of production in Latin America

Problems and challenges for trade unions in the cases of Argentina, Mexico, and Chile

Victoria Basualdo, Graciela Bensusán, and Dasten Julián-Vejar

Introduction

In line with the changes taking place to the global pattern of capital accumulation, we are presently witnessing a new industrial and/or technological revolution that guarantees to transform social relations and communication, work, and everyday life. This fact, which has been widely commented on by the literature at the international level, has opened up an interesting debate that, through measurements, projections, and speculations, highlights such processes as technological unemployment, the disqualification of work, the automation of various tasks and processes, and the robotization of certain jobs.

However, these processes take on their own expressions and particularities in terms of the reality of Latin America and the Caribbean, as they occur gradually, are differentiated and are limited by the characteristics of companies, productive sectors, geographic locations, state policies, business projects, and labor relations. Despite these particularities, there are also similarities with the processes witnessed at the global level, especially in the case of the implications for the organization of work, transformations in the processes of labor, and the problems generated for the structure of employment.

At present, there is an important and emerging literature focused on the technological changes introduced by the 4.0 revolution in the Latin American region, which involve analysis undertaken by multilateral bodies such as ECLAC, the ILO, and OECD, as well as important references in studies on labor, engineering, economics, and public policy. Within the current technological revolution, one of the problems that has already been addressed at the global level, and that presents its own peculiarities in Latin America, is related to labor relations and the current challenges for trade unions.

The power of organizations, their repertoires of action, their models of negotiation and their capacity to influence various levels of labor relations, generate a necessary reflection on the possibilities of harmonizing technological changes in relation to the needs and demands of workers. This becomes even more necessary taking into account the realities of inequality, discrimination, and poverty in the region, especially if we consider the potentially negative effects of the introduction of processes of automation and/or robotization, such as technological unemployment and the polarization of the labor market.

In this chapter, we will present a review of some of the debates regarding automation, robotization, and the uses of artificial intelligence in Latin America and the Caribbean. We will focus on providing an account of the problems that the 4.0 revolution involves for trade union organizations, and the nature of the main opportunities that are emerging from this new scenario. This work will be exemplified and expanded in the cases of Argentina, Mexico, and Chile. Finally, we will present certain conclusions and perspectives for organizations, taking into account the social protests that have occurred and the new correlation of political forces in the region.

State of affairs

As can be seen in the abundant literature available, throughout history technological revolutions have been carried out and have happened, that dramatically transformed production and labor processes, and that these changes have represented fundamental pillars for the expansion and reproduction of capitalism. It is generally accepted that technological change is a non-linear, evolutionary, and resource-intensive process that is driven not only by economic forces, but also by political and social forces. It is not a homogeneous process and generally involves different forms of change and innovation that can affect the quantity and character of labor tasks in very different ways (Nübler, 2016). Technological change is thus reflected in the generation of new knowledge, the implementation of new forms of production, and in the modification of the products obtained in labor processes, as well as in new production techniques, changes in the organization in the work place, and modifications in terms of the diffusion of these changes in the economy as a whole.

This is possible to verify in the history of the world economy, based on certain nuances and distinctions, which have different expressions depending on geographical scales, geopolitical positions, and the role of the state in the process of capital accumulation. At present, the novelty of the technological changes taking place recognizes the emergence of a 4.0 Industry, one that represents the basis of a new technological revolution.

Klaus Schwab (2016) refers to this time as "the fourth industrial revolution," which would not only consist of interconnected intelligent machines and systems, but by the integration of multiple waves of technological progress "ranging from genetic sequencing to nanotechnology, and from renewable energy to quantum computing." For this reason, the present revolution may be characterized by "the fusion of these technologies and their interaction through the physical, digital and biological domains," which makes this revolution fundamentally different to all those that proceeded it.

At the same time, it is also a revolution "where emerging technologies and broad-based innovation are spreading much faster and more widely than previous revolutions, some of which are still underway in some parts of the world" (Schwab, 2016, p. 21). This revolution has been accompanied by the processes of automation and digitization of production. In this process the incorporation of new technologies occurs in the undertaking of tasks that previously required higher cognitive levels, and which are characterized by being not very repetitive, and supposedly to be exclusively undertaken by people, even those with a technical or professional background. Added to this, their speed and impact on various productive sectors (retail, the automobile sector, transport, telecommunications, etc.), generate important changes in the world of work.

The process of automation has taken on new characteristics, which challenge the categories and conceptualizations of industrial and labor sociology. This form of automation involves a combination and escalation of the capabilities developed and available from the digitization process of production, delivering new forms of technological replacement of work through the programming, exercise, and operation of artificial intelligence.

Artificial intelligence includes a set of applications and manifestations that can be translated into various technological devices, platforms, supports, and hardware that can be introduced into everyday working life, such as in production chains (Ernst, Merola & Samaan, 2018; Aghion, Antonin & Bunel, 2019). One of the forms manifested by this process in the 4.0 Revolution is robotization. Although this particular process is often viewed as being separate from digitization and automation, robotization incorporates and is part of the fusion of various forms of technology.

However, and although it tends to be understood as a homogenizing and optimistic discourse on technological change, it is necessary to recognize that this revolution has been created thanks to the actions of capital and business and disseminating mainly from the most industrially developed nations. This fact involves two important consequences when considering the reality of work in Latin America and the Caribbean in relation to this ongoing process:

1 There is a deficit in terms of the analysis of technological change beyond unemployment and productive restructuring, particularly when dealing with the role, interest, action, incidence, resistance, consent, and organization of workers in this process; and

2 The reality of the continent's technological and industrial development suggests a phenomenon that can be located only in specific and limited sectors, rather than as a generalized and extensive process incorporating the whole labor market.

Although the academic literature at a global level has verified that with this revolution new jobs have been created (for example, the delivery of food or goods), while other activities have been modernized, thus destabilizing traditional professions such as taxi drivers, tour guides, and travel agents, etc., there is a need to revisit contributions based on the realities of specific and situated territories.

These contributions are based on the analysis of trajectories and experiences of productive and/or technological changes that differ from those offered by industrialized countries and by corporate discourse. This focus offers the possibility of simultaneously recovering the perspective of working subjects and their union organizations, in terms of a contextualized and historical perspective.

In this sense, the specificity of Latin America and the Caribbean in this process of technological change is consistent with its articulation with the following:

1 Phenomena that have straddled labor relations with a particular intensity and importance since the 1970s, highlighting their intensification in the 1990s, through labor outsourcing and the so-called productive restructuring that imposed very significant dynamics in terms of the erosion of corporate responsibility and profound transformations of labor relations (Basualdo & Morales, 2014; Basualdo, Esponda, Gianibelli & Morales, 2015; Etchemendy, 2018). This process was characterized by the introduction of neoliberal policies, the persistence of an unregistered labor sector, and the escalating precariousness of the labor market (Julián, 2017).

2 Regressive labor reform processes that have sought to be imposed, with various formats and nuances (Bensusán, 2013; Basualdo, 2019), but which share the common characteristic of a motivation to launch an attack against a structure of labor rights that were fought for during decades of organization and the struggle of workers' organizations. All these processes, dynamics, and disputes, had and have an impact on individuals and labor and union organizations, leading to their weakening and fragmentation.

Taking into account these two levels, it is necessary to introduce into this debate the reality and action of workers, considering the forms that the resistances, struggles and responses of adaptation, negotiation and conflict have taken in the processes of technological revolution. The reality of workers does not tend to be considered in these debates and, in the case of Latin America and the Caribbean, we are referring to very complex national, sectoral, and occupational situations. Consequently, we will try to account for three diverse cases in geographic, economic, and socio-political terms.

The case of Argentina

In the case of Argentina, which is characterized by a relatively high rate of union affiliation and the existence of strong domestic union bodies organized by branches of activity and articulated in national union confederations, there has been an extensive debate and academic output regarding technological change and innovation in the country, and their impact on labor relations. This production increased notably in recent years in connection with the development of the agenda proposed by the ILO regarding the "Future of work," and in dialogue with various international lines of investigation (Ernst & Robert, 2019; Catalano, 2018; Bortz, Moncaut, Robert, Sarabia & Vázquez, 2018, among many others). In this context, special attention was paid both to the latest transformations of the technological frontier, and to the appearance of new activities, particularly the platform economy and its challenge in terms of labor laws, union organization, and state oversight in a context of precarious working conditions (Roiter, 2019; Scasserra 2018 & 2019; Del Bono, 2019; Madariaga, Buenadicha, Molina & Ernst, 2019). Even more challenging is the reception of these research contributions by the trade union movement, the agenda of which is strongly marked by fragmentation and increasing job insecurity as a legacy of a cycle of neoliberal reforms, and by the adjustment and various attempts at regressive labor reforms carried out between 2015 and 2019.

Recent contributions have allowed the presence and influence of technological change in industrial, agro-industrial, extractive, and service activities to be revealed. An investigation carried out by an interdisciplinary team coordinated by the Economics and Technology Area of the Latin American Faculty of Social Sciences (FLACSO), with the support of the Friedrich Ebert Foundation (Basualdo, et al., 2019), involved a primary analysis of responses and union positions both regarding the processes of technological change, and in relation to labor outsourcing in six relevant economic activities in Argentina: the steel industry, the automobile and sugar industries, the oil industry, and two service activities: telephony and banking, which partially includes public sector employment. This occurred in the 2003–2018 period.[1]

Case analysis showed that technological change constitutes a central phenomenon in all the economic activities analyzed, and that therefore it is not appropriate to think of it as an issue of "future work," but rather as one with a clear presence and a close link with work being undertaken right now. Thus, this must also be analyzed within a context of vigorous attempts at regressive labor reform based on a set of labor and economic public policies.

Collective bargaining appeared in this survey as a key area, and it is possible to distinguish, in the period addressed, two substantially different stages in Argentina. Between 2003 and 2015 there was a boom and growth in collective bargaining, even with its prevalent tensions and contradictions, and within a context of growth in the industrial sector and the economy as a whole, as well as a progressive redistribution of income. Since 2016, there has been a growing stagnation, and within the context of a record drop in industrial production, a rise in inflation and regressive state interventions, in addition to a significant decline in labor rights in various areas. This became manifest in 2018 in the elimination of the Ministry of Labor and its replacement by a Secretariat.

An analysis of the presence of these issues in collective bargaining made it possible to register that this area is key for employers as well as unions and workers. With respect to trade union organizations, the presence of clauses establishing the need to preserve sources of employment in the event of technological change and to guarantee the training and education of workers, underlined the concern about the possible impact of technological changes on the level and conditions of employment. On the part of business sectors, cases such as that of the oil industry exemplified how the use of processes of technological change, and the opening up of new activities, were a cause for the introduction of clauses in collective bargaining that imply the loss of basic labor rights. In this case, it can also be observed that the possible impact of technological change is not only the direct loss of jobs but can also translate into a significant loss of rights, constituting another way to introduce regressive "labor reforms."

In all cases it was concluded that in order to understand the process, it was necessary to include other aspects in dialogue with collective bargaining. One of these was the process of conflict with respect to technological change and/or labor outsourcing, in various activities such as banking, the steel industry, and telephony (particularly regarding labor outsourcing), while to a lesser extent, significant conflicts were registered in the sugar industry. However, no visible and general conflicts were identified in the oil and car industries. New initiatives and instances of union organization were also highlighted, such as the creation of union networks by a group of companies, particularly in the steel industry, in the face of the changes that had appeared in forms of work and production.

With respect to union institutions that study these problems, an experience was recorded of telephone companies forming a Secretariat of New Technologies, as well as the formation of a Telecommunications Research and Development Institute, which were viewed as interesting alternatives. At the same time, the development of trade union training initiatives closely linked to technological changes, drew a lot of positive attention in industrial and services cases. Finally, another key area in order to study both issues would seem to be that of the Safety and Hygiene Committees, which were established by law in the provinces of Buenos Aires and Santa Fe. In several of the activities, particularly in industries such as the steel or car sectors, the committees appeared as useful instances to deal with issues related to working conditions and health impacts, containing and dealing with issues of technological change and outsourcing, and highlighting the different conditions that this created between permanent and outsourced workers.

These lines of union organization and action in defense of labor rights and working, living, and organizational conditions, occurred during a stage of robust technological change and transformation of labor relations, marked by the tension and debate on business policies, and with strong pressures and contradictions within the union movement itself. This also included dealing with highly regressive state policies, particularly between the end of 2015 and 2019. Even though these initiatives were in many cases fragmentary and disparate, they underline the existence of valuable and frequently obscured processes of organizing workers and unions that would be important to analyze in greater depth.

The case of Mexico

An important line of research on technological change and its implications in Mexico focuses on identifying the factors that can accelerate or reduce its impact on employment, as well as the sectors, occupations, and jobs that are being most affected (Rodríguez Pérez, Camberos Castro & Huesca Reynoso, 2015; Huesca & Ochoa, 2016). The problems related to the regulation of jobs in emerging occupations have also been studied, such as those generated by digital platforms (Bensusán, 2017 & 2020).

In the union sector, the weakened situation of organizations after decades of deterioration due to corporate arrangements that ceased to create favorable conditions for workers, and generated an extremely feigned approach to collective bargaining, has led to the predominance of defensive positions regarding the consequences of technological innovation and the organization of work in employment. There has been scant interest regarding unions in research on the effects of new technologies and productive restructuring, barring a few exceptions. However, an issue that has recently been the focus of attention has been the accelerated expansion of subcontracting processes, with negative consequences for workers. Along with other labor reforms adopted in recent years at the constitutional (2017) and legal (2019) levels, which created favorable conditions for the democratization of union organizations and the strengthening of collective bargaining, unions have been demanding the adoption of a more restrictive regulation to stop outsourcing employment, especially as this is associated with job instability, tax evasion, and social security.

Technological innovation is associated with integration in North America through NAFTA (1994), which was renegotiated in 2018 and renamed as the USMCA (entering into force on July 1, 2020). This process gave rise to an export model based on low labor costs, increasing productivity, and proximity to consumers. Huesca and Ochoa (2016, p. 168) argue that the economic restructuring, the increase in foreign direct investment, and the maquiladora system favor the insertion of multinational companies. These processes have increased the use of technology, generating changes in the demand for labor and wages, with greater intensity in the manufacturing industry. However, they consider that technological change was essentially dependent on the decisions taken by the corporations in the countries of origin of the maquila factories, and which had invested in Mexico attracted by low labor costs, which delayed the progress of innovation. One exception, however, was that of the car industry, where such processes have been accelerated.

In this context, interest was generated in academia to ascertain the relationship between these processes and the opportunities or restrictions to improve the quality of jobs in multinational companies. A study carried out by an inter-institutional and interdisciplinary team focused on the experiences of multinational companies with successful trajectories located in five sectors, where innovation combined, in each case, the use of new technologies, and the reorganization of work and business models employed.[2] One result of this research was the fragility, insufficiency or absence of multiplier effects of innovation at the level of wages and working conditions, the participation of workers in the company or in unions (democracy and labor representation), or in workers' professional development (further on-the-job training). Among the most important factors in explaining such insignificant benefits obtained by workers, were problems in the quality and structure of union representation, with the consequent fragmentation and distortion of collective bargaining. Another negative influence was the context in which such innovation processes occurred, characterized by an extremely restrictive wages policy that drove as a consequence greater labor flexibility, although this began to be reversed after the arrival of a new government in December 2018.

The case of the car industry, where all assembly plants have unions and the permanent workforce is unionized, illustrates this issue. More than to obtain advantages associated with the success of the export model – the sector has become the seventh largest car producer in the world – collective bargaining in assembly plants was an efficient instrument to ensure the absence of strikes in a context of increasing concessions to multinational subsidiaries. The process of automation and robotization was accompanied by job losses in the assembly plants, while there were increases in auto parts companies. There has also been greater labor and salary flexibility, and increasing segmentation of working conditions according to length of employment and workers' places in company value chains. In contrast, the few instances of conflict, in which workers sought to dispute representation with the unions as a condition for regaining their bargaining power, ended in defeats, including layoffs, company closures, and production being shifted to other countries (Bensusán & Covarrubias, 2016; Pardi et al., 2018).[3]

It should be noted that out of a total of 18 plants belonging to the Mexican car industry, only in the case of VWM, along with Nissan Cuernavaca and Audi, are there independent and active unions, which following major conflicts and ruptures in the late 1980s and early 1990s, managed to sustain an important capacity for dialogue with their respective companies. However, the limits of this arrangement lie, among other factors, in the isolation of such organizations, given the undemocratic and subordinate characteristics prevailing in the rest of the union movement in the car sector and the difficulties in making any progress in the formation of broader coalitions (Bensusán & Gómez, 2017).

However, and to conclude on a more optimistic note, it can be said that the proactive salary policy of the country's new administration (2018–2024), the existence of new rules of the game in the area of collective rights, and the requirements imposed by USMCA, which references the implementation of said rules, may create in the immediate future a more favorable context for unions and collective bargaining. Consequently, and unlike what occurred in the 1990s, USMCA may bring more favorable conditions to achieve a positive articulation between the progress of technological change and improvement of job quality, particularly in terms of wages.

The case of Chile

Regarding the case of Chile, we can summaries some of the results that we have been able to obtain from the Fondecyt Regular 1161347 project "Mapping of precarious work in the southern Macro Zone" (2016–2019), where the new and old conditions of precariousness are discussed from the point of view of the collective action and associativity of workers. In this project, it was possible to verify some of the occupational profiles taken on, rejected, and required in the process of automation, digitization, and robotization of work, from the point of view of the labor market.

In Chile, the automation process is framed within a government and social model characterized by the application of neoliberal policies, a deliberate weakening of union organizations, and a process of social and job insecurity. The country's economic model is focused on the export of raw materials, from which emerge the country's main productive sectors: the mining industry, the forestry, agricultural, and aquaculture sectors. Each of these has union organizations with permanent workers and a large chain of subcontracted workers. The country's productive matrix is made up of these primary or commodity production sectors, which are part of the economic pattern of low-level industrialization and minimum investment in research, technology, and development.

Studies on the current process of automation and robotization are in their early stages in Chile (Almeida et al., 2017; Bravo, et al., 2018; Carrillo, 2018; Rivera, 2019). However, research can be found that indirectly, places said process as one of the keys to understanding the world of contemporary labor. Research aimed at this issue is becoming increasingly popular, and combines the relationship between traditional sectors that aim at automation and robotization, as well as the emergence of jobs from so-called platform capitalism (Uber, Orders Now, etc.) (Bravo et al., 2019).

Through various investigations, we were able to account for how the automation process has threatened traditional union organizations, which have a structural and organizational power of greater dimension than smaller and younger unions (Julián, 2017; 2018; 2019). This has emerged mainly in the case of mining, transport, and the retail sector, where the replacement by means of digitizing machines or price readers, and the operation of intelligent logistics systems, and artificial intelligence in the management of trucks, has become a major threat to jobs.

To this we must add the fragility with which unions have been politically instituted in Chile, due to regulatory frameworks that have weakened the role of organizations in dealing with negotiations and strike processes, their funding sources and decentralization at the company level, as well as the issues that can be negotiated with the employer. Among the restrictions on those issues that can be negotiated collectively, are changes at the productive and/or organizational level of the company, including technological changes. Only a few unions manage to introduce resistance to vertical processes of labor automation through the use of various repertoires of collective action.

In general, we studied organizations that have a reactive logic of defense, which has an extremely minimal corrective effect in relation to business strategies. The 2017 labor reform that modifies some of the collective rights of union organizations (Julián, 2019), did not include matters of collective bargaining, which has resulted in a serious problem in terms of conflict resolution and dialogue with companies. Moreover, there has been no interest or union culture able to internalize and incorporate practices of research, information, and public debate regarding technological change and the replacement of workers.

However, despite this trend, there are also experiences of unions that have tried to organize campaigns and protests against automation processes. Some of these unions are in the mining sector, in Santiago's subway train (metro) system, in the banking and retail sectors, and logistics, among others. However, processes of technological change are generally introduced without consultation with or participation of the trade unions. The orientation of automation involves the strategy of:1) replacement of workers; 2) introduction of multitasking; and 3) the weakening of established trade union organizations.

The responses that the unions have developed to these processes are mainly related to the reaffirmation of a legal-institutional strategy, consisting of challenging such processes in labor courts and legal conflicts related to dismissals. Furthermore, there are collective bargaining processes that include clauses aimed at the retraining of workers, through an orientation towards reskilling in terms of the functions of the company and/or the dismissal of workers. Both processes are marked by the presence of instances of qualification training designed for self-employment (plumbers, electricians, etc.) and/or for new activities within the company. This represents the restructuring of production.

One of the exceptions to this set of issues has been that of Walmart Chile. As part of a collective bargaining process, the Leader Inter-Company Workers' Union of Walmart Chile, established the process of job automation as an issue requiring negotiation. After suffering a series of layoffs, the battle of the trade union organizations was directed towards the protection of employment and labor rights. This led to a union mobilization incorporating 17 thousand workers, who, within the framework of exercising their legal right to strike, were able to negotiate the issues of restructuring and dismissal.

On processing the forms of multitasking as forms of productive restructuring, along with dismissals, work overloads, and the lack of guarantees in employment, precarious labor forms a central part in understanding the automation process in Chile. While business strategies refer to new restructuring processes, unions have begun to make it a part of their new repertoires of action, placing it in relation to the context of induced precariousness in the labor market.

Conclusions

In addition to the fact that the progress of technological change has been uneven in the region and, in particular, in the countries studied and within them, according to the economic sectors, it is necessary to consider the marked differences in the power resources (Schmalz, 2017) that are available to trade unions to counter the worst effects of automation and robotization on the quantity and quality of jobs. While unions in Argentina have historically had sectoral unions and collective bargaining structures with high levels of union affiliation and bargaining coverage, Mexico and Chile are characterized by a growing weakening of the trade union movement. For different reasons – historical, institutional, and economic – the structural and instrumental power of employers to impose their conditions is much greater in these two countries than in Argentina.

In this sense, and if some capacity for trade union dialogue can be expected in the face of the ongoing innovation processes anywhere in the continent, that would be Argentina, where along with traditional power resources can be added the growing interest of the unions in the subject matter under analysis, together with the presence, since the end of 2019, of a government with similar interests to those of the workers. Albeit that the country's government is now facing major constraints due to the levels of external debt inherited from the previous government and the unprecedented crisis created by the COVID-19 pandemic. In the current critical situation, the processes of labor restructuring and technological changes, the force of business pressures that seek to condition state policies, and the very tensions and limitations of union organizations, some of which have taken defensive or even concessional positions, have all come into play. Consequently, to reverse these labor and union impediments, it is essential to recover the experiences of trade unions in the defense of labor rights.

In the case of Mexico, where there is also an administration favorable to the creation of counterweights to the power of big business, the recent labor reform has opened an exceptional opportunity to transform the poor quality of representation in the immediate future and expand experiences of authentic dialogue. This is a possibility marked by uncertainty, given the difficulty of overcoming the enormous loss of prestige of unions due to decades of simulation in the exercise of collective rights.

With respect to Chile, the social and popular rebellion that began in October 2019 has not yet lost its power to instigate changes in the political system and in labor relations. However, in the context of the COVID-19 pandemic emergency, the radicalization of the labor programmeof the right-wing neoliberal coalition presently in government, points towards a weakening of trade union organizations and their possibilities to negotiate or resist processes of automation and robotization in the workplace. Added to this is the fact that the difficulties present for the development of union repertoires are increased by the strength and cohesion of the business and financial class to impose their interests, given the country's political system and the way government is exercised.

As it is possible to see, technological change takes on a range of national expressions in Latin America, in terms of societies in a state of broad transformation and change, and with a potential economic crisis in the making that threatens the sources of employment for millions of workers. The asymmetries between capital and labor, between social classes and their organizations, and the institutional and cultural particularities and differences in the geographical scales of capitalism, present a diverse and contradictory panorama when trying to ascertain the present and future of technological change in Latin America.

Notes

1 This research was also carried out within the framework of the Scientific and Technological Research Projects (PICT) 2016-0575: "Structural transformations and labour relations: productive changes and union strategies in the steel and sugar industry from the mid-1970s to the present day" and received comments, suggestions and feedback from academic experts and union leaders from the sectors involved.

2 The investigation, which was carried out between 2013 and 2015, included a total of 16 multinational companies in the car sector (5); aerospace (3), electronics (4) and services (4). Their results were published in Carrillo, Bensusán and Micheli, 2017.

3 On labor relations in the Mexican car industry see also De la Garza Toledo & Hérnandez Romo, 2018.

References

Aghion, P., Antonin, C., & Bunel, S. (2019). Artificial Intelligence, Growth and Employment: The Role of Policy, *Revista Economie et Statistique*, 510–512, 149–164. doi: 10.24187/ecostat.2019.510t.1994

Almeida, R. K., Fernándes, A. M., & Viollaz, M. (2017). *Does the adoption of complex software impact employment composition and the skill content of occupations? Evidence from Chilean firms* (Working Paper No. 8110). Washington, D.C.: World Bank.

Basualdo, V., & Morales, D. (2014). *La tercerización laboral. Orígenes, impacto y claves para su análisis en América Latina [Labor outsourcing. Origins, impact and keys for its analysis in Latin America].* Buenos Aires: Editorial Siglo Veintiuno.

Basualdo, V., Esponda, A., Gianibelli, G. & Morales, D. (2015). *Tercerización y derechos laborales en la Argentina actual. [Outsourcing and labor rights in Argentina today]* Buenos Aires: Editorial de la Universidad Nacional de Quilmes-Página 12.

Basualdo, V. (2019). Los intentos de Reforma Laboral regresiva en la Argentina desde 2015: una lectura en perspectiva histórica. [Regressive Labor Reform attempts in Argentina since 2015: a reading in historical perspective] In C. G. Speranza (org.), *História do Trabalho. Entre debates, caminhos e encruzilhadas*, São Paulo: Paco Editorial.

Basualdo, V., Letcher, H., Nassif, S., Barrera, M., Bosch, N., Copani, A., Rojas, M. (2019). *Cambio tecnológico, tercerización laboral e impactos sobre el empleo. Desafíos desde y para una narrativa argentina. [Technological change, labor outsourcing and impacts on employment. Challenges from and for an Argentine narrative].* Buenos Aires: Fundación Friedrich Ebert. Retrieved from: https://www.fes-argentina.org/

Bensusán, G. (2013). Labour Reform from a Regional Perspectives: Experiences in the Americas. In A. Blackett and C. Levésque (Eds.), *Social Regionalism in the Global Economy* (pp. 207–224). New York & London: Routladge,

Bensusán, G., & Covarrubias, A. (2016). Relaciones laborales y salariales en la Industria Automotriz: ¿podrá el cambio venir de afuera?, *[Wage and labor relations in the Automotive Industry: Will change come from outside?]* In A. Covarrubias, S. Sandoval, G. Bensusán & A. Arteaga (Eds.), *La industria automotriz en México, Relaciones de Empleo, Culturas Laborales y Factores Psicosociales.* Hermosillo: El Colegio de Sonora-Red ITIAM.

Bensusán, G. (2017). Nuevas tendencias en el empleo: retos y opciones para las regulaciones y políticas de mercado de trabajo. [New trends in employment: challenges and options for labor market regulations and policies]. In G. Bensusán, W. Eichorst & J. M. Rodríguez (Eds.), *Las tecnologías y sus desafíos en el empleo, las relaciones laborales y la identificación de las demandas de cualificaciones [Technologies and their challenges in employment, labor relations and the identification of qualification demands]* (pp. 89–171), Santiago de Chile: CEPAL. Retrieved from https://www.cepal.org/es/publicaciones/42539-transformaciones-tecnologicas-sus-desafios-empleo-relaciones-laborales-la

Bensusán, G. & Gómez W. (2017), Volkswagen de México: un caso de articulación positiva y fuerte pero inestable entre innovación productiva y social. [Volkswagen de México: a case of positive and strong but unstable articulation between productive and social innovation]. In J. Carrillo, G. Bensusán & J. Micheli (Eds.), *¿Es posible innovar y mejorar laboralmente? Estudio de trayectorias de empresas multinacionales en México* (pp. 183–238), México: UAM-A.

Bensusán, G. (2020). *Ocupaciones emergentes en la economía digital y su regulación en México, [Emerging occupations in the digital economy and its regulation in Mexico], Cepal, Santiago de Chile.* Santiago de Chile: CEPAL. Retrieved from https://www.cepal.org/es/publicaciones/45481-ocupaciones-emergentes-la-economia-digital-su-regulacion-mexico

Bortz, P., Moncaut, N., Robert, V., Sarabia, M. & Vázquez, D. (2018). *Cambios tecnológicos, laborales y exigencias de formación profesional. Marco y dinámica institucional para el desarrollo de las habilidades colectivas. [Technological changes, labor and professional training requirements. Institutional framework and dynamics for the development of collective skills] (Workink Paper No. 23).* Buenos Aires: OIT. Retrieved from: https://www.ilo.org/wcmsp5/groups/public/—americas/—ro-lima/—ilo-buenos_aires/documents/publication/wcms_643791.pdf

Bravo, J., García, M. A., & Schlechter, H. (2018). *Automatización e Inteligencia Artificial, desafíos del Mercado Laboral [Automation and Artificial Intelligence, challenges of the Labor Market]* (Working Paper No. 50). Santiago de Chile: CLAPES UC. Retrieved from https://clapesuc.cl/assets/uploads/2018/10/15-10-18-doc-de-trabajo-automatizacion-vf-2018.pdf

Bravo, J., García, M. A., & Schlechter, H. (2019). *Mercado Laboral Chileno para la Cuarta Revolución Industrial [Chilean Labor Market for the Fourth Industrial Revolution]* (Working Paper No. 59). Santiago de Chile: CLAPES UC. Retrieved from https://clapesuc.cl/investigaciones/doc-trabajo-no59-mercado-laboral-chileno-para-la-cuarta-revolucion-industrial/

Camberos Castro, M., & Huesca Reynoso, L. (2015). *Mercado laboral y cambio tecnológico en México: Tendencias, sectores y regiones. [Labor market and technological change in Mexico: Trends, sectors and regions]* México: CIAD-LIBERMEX.

Carrillo, F. (2018). *Formación de Competencias para el Trabajo en Chile. Comisión Nacional de Productividad. [Skills training for Work in Chile. National Productivity Commission]. Retrieved from* https://www.comisiondeproductividad.cl/wp-content/uploads/2018/03/Informe_Formacion-de_Competencias-para_el_Trabajo.pdf

Catalano, A. (2018), *Tecnología, innovación y competencias ocupacionales en la sociedad del conocimiento [Technology, innovation and occupational skills in the knowledge society]* (Working Paper No. 22). Buenos Aires, OIT. Retrieved from https://www.ilo.org/wcmsp5/groups/public/—americas/—ro-lima/—ilo-buenos_aires/documents/publication/wcms_635946.pdf

De la Garza Toledo, E., & Hérnandez Romo, M. (2018). *Configuraciones productivas y laborales en la tercera generación de la industria automotriz en México.* México: Miguel Ángel Porrúa / UAM.

Del Bono, A. (2019). Trabajadores de Plataformas Digitales: condiciones laborales en plataformas de reparto a domicilio en Argentina. *[Digital Platform Workers: working conditions on home delivery platforms] Revista Cuestiones de Sociología,* 21, 1–30. doi: https://doi.org/10.24215/23468904e083

Ernst, C., & Robert, V. (2019). *Cambio tecnológico y futuro del trabajo. Competencias laborales y habilidades colectivas para una nueva matriz productiva en Argentina. [Technological change and future of work. Labor competencies and collective skills for a new productive matrix in Argentina]* Buenos Aires: OIT. Retrieved from https://www.ilo.org/buenosaires/publicaciones/documentos-de-trabajo/WCMS_734829/lang–es/index.htm

Ernst, E., Merola, R., & Samaan, D. (2018). *The economics of artificial intelligence: Implications for the future of work*, Geneva: ILO. Retrieved from https://www.ilo.org/wcmsp5/groups/public/—dgreports/—cabinet/documents/publication/wcms_647306.pdf

Etchemendy, S. (2018). La tercerización laboral en la Argentina. Diagnóstico y estrategias sindicales. [Labor outsourcing in Argentina. Union diagnosis and strategies] Buenos Aires: Editorial Biblos.

Huesca, L. & Ochoa, G. (2016). Desigualdad Salarial y Cambio Tecnológico en la Frontera Norte de México.[Wage Inequality and Technological Change in the Northern Border of Mexico] *Problemas del Desarrollo*, 47(187), 165–188. doi: https://doi.org/10.1016/j.rpd.2016.10.006

Julián, D. (2017). Unions Opposing Labor Precarity in Chile. Union Leaders's Perceptions and Representations of Collective Action. *Latin American Perspectives*. 45 (1), 63–76. doi: https://doi.org/10.1177/0094582X17737494

Julián, D. (2018). Precariedad laboral y repertorios sindicales en el neoliberalismo. Cambios en la politización del trabajo en Chile. [Job insecurity and union repertoires in neoliberalism. Changes in the politicization of work in Chile] *Revista Psicoperspectivas*. 17(1). doi: https://doi.org/10.5027/psicoperspectivas-vol17-issue1-fulltext-947

Julián, D. (2019). Transformación y bifurcación de las trayectorias sindicales en Chile: La "reforma laboral" y el escenario sindical (2014 – 2016). [Transformation and bifurcation of union trajectories in Chile: The "labor reform" and the union scenario (2014 - 2016)] *Revista Izquierdas*, 49, 1696–1714. Retrieved from https://www.academia.edu/41191953/Transformacion_y_bifurcacion_de_las_trayectorias_sindicales_en_Chile

Madariaga, J., Buenadicha, C., Molina, E., & Ernst, C. (2019), *Economía de Plataformas y Empleo. ¿Cómo es trabajar para una app en Argentina? [Platform Economy and Employment. What is it like to work for an app in Argentina?]* Buenos Aires: CIPPEC-BID-OIT.

Nübler, I. (2016). *New technologies: a jobless future of golden age of job creation* (Working paper No. 13). Geneva: ILO. Retrieved from https://www.ilo.org/global/research/publications/working-papers/WCMS_544189/lang–en/index.htm

Pardi, T., Dicer, E., Wenten, F., Carrillo, J., D'Costa AP., Krzywdzinski, M., … Lüthje, B. (2018). The Future of work in the automotive industry II. *Strategies, Technologies and Institutions.* Paris: Gerpisa.

Rivera, T. (2019). Efectos de la automatización en el empleo en Chile. *Revista de análisis económico*, [Effects of automation on employment in Chile. Economic analysis magazine], 34(1), 3–49. doi: https://dx.doi.org/10.4067/S0718-88702019000100003

Roiter, S. (2019). *Cambio tecnológico y empleo: aportes conceptuales y evidencia frente a la dinámica en curso [Technological change and employment: conceptual contributions and evidence against the current dynamics]* (Working Paper No. 15.1). Buenos Aires: CIECTI. Retrieved from http://www.ciecti.org.ar/wp-content/uploads/2019/01/DT15.1_v2.pdf

Scasserra, S. (2018). *Las plataformas web (y qué demandar desde el sindicalismo lati-noamericano). [Web platforms (and what to demand from Latin American union-ism)]* Buenos Aires: Fundación Friedrich Ebert.

Scasserra, S. (2019). El despotismo de los algoritmos. Cómo regular el empleo en las plataformas. [The despotism of algorithms. How to regulate employment on platforms] *Revista Nueva Sociedad*, 279, 133–140. Retrieved from https://dialnet.unirioja.es/servlet/articulo?codigo=7005190

Schmalz, S. (2017). Los recursos de poder para la transformación sindical. [Power resources for union transformation] *Revista Nueva Sociedad*, 272, 19–41. Retrieved from https://dialnet.unirioja.es/servlet/articulo?codigo=6202156

Schwab, K. (2016). *La cuarta revolución industrial. [The fourth industrial revolution]* Barcelona: Debate.

10 Using functional and social robots to help during the COVID-19 pandemic

Looking into the incipient case of Chile and its future artificial intelligence policy

Carmina Rodríguez-Hidalgo

Flying robots: when help arrives from the sky

"We have now a solution to help the high-risk population to obtain medicines, without exposing a public worker or a family member to the person that is in quarantine," tells mayor Gustavo Alessandri regarding a pilot drone assistant deployed in April 2020 in the Chilean coastal town of Zapallar, some 2.5 hours away from Santiago, Chile's capital. Zapallar is one of the most exclusive areas, but also one where great social inequalities persist. The drones are being used to keep some of its poorest and oldest inhabitants, who often reside in areas difficult to access, out of danger of contagion (Núñez-Torrón, April 23, 2020). In particular, the pilot was deployed in the areas of Zapallar of La Hacienda, El Blanquillo, Población Estadio, and La Retamilla. Apart from medicines, the drone delivers protective supplies to combat the virus, such as disinfectant alcohol gel, gloves, and face masks. "All the equipment and infrastructure in our municipality are at the service of our neighbors to face this sanitary emergency," tells major Alessandri to the national TV news (24horas, April 16, 2020). Press footage shows the drone high in the sky carrying a white bag, then landing in a poor district where two elderly people wearing face masks pick up the bag and look inside to gather the supplies (Cooperativa, April 17, 2020).

The type of drone seen in the footage is a quadcopter, which falls under the category of "flying mobile robots." This drone has the potential of supporting and improving the quality and availability of health care in challenging, emergency contexts such as a pandemic. In fact, drone technology has been highlighted as one of the ten technologies to fight Coronavirus, in a recent report by the Scientific Foresight Unit (STOA) of the EPRS, the European Parliamentary Research Service (EPRS, April 2020). Just like in the example of the assisting drone in Zapallar, the report recognizes the drones' potential to "facilitate the tasks of enforcing containment and social distancing measures, helping reduce the number of face-to-face contacts, but also freeing up crucial human resources (such as health workers and law enforcement officers), while minimizing exposure to the virus, thereby reducing the chances of contamination" (*p.* 16). Moreover, according to Tavakoli, Carriere & Torabi (2020), robotic technologies such as flying drones bring forth "the practical and life-saving benefits from incorporating robotics and automation technologies into the health care system. While automation is often depicted in popular culture as a force that eliminates jobs, now more than ever, we need to consider its life-savings potentials as well" (*p.* 2).

Ethical challenges of drones

In spite of these advantages, the EPRS report highlights caveats and recommendations for anticipatory policymaking regarding the use of drones during a pandemic. Because this type of drone collects data and their deployment may be hastened due to an emergency context, "specific safeguards need to be introduced so that full protections are afforded to personal data once the state of the emergency is lifted" (EPRS, *p.* 18). The report cites the European Data Protection Board (EBPD) on the processing of personal data in the context of the COVID-19 outbreak. This board advised public authorities against "systematic and generalized" monitoring and collection of health-related data, recommending the processing of location data in an anonymous way. However, the chair of the EDPB clarified that

> safeguarding public health may fall under the national and/or public security exception of Article 15 of the Directive, which enables the Member States to introduce legislative measures pursuing national security and public security. Although many of the exceptional measures controlling the use of drones are based on extraordinary powers, only to be used temporarily in emergencies, specific safeguards need to be introduced so that full protections are afforded to personal data once the state of emergency is lifted (*p.* 18).

In Chile, a country which at the time of writing is developing its national artificial intelligence policy, there is concern for the ethical implications of AI and the privacy of data, particularly location data. Privacy and the protection of personal data are subpillars of the country's AI policy, which is being discussed and formed at the time of writing. "Data are a basic enabling condition for every AI system," reads an internal report of the Ministry of Science, Technology, Knowledge, and Innovation (Equipo Futuro, 2020). "Data is actually the building block of AI. Without it, we don't have anything," tells José Guridi, advisor to Chile's Ministry of Science.

The robot is the message: a social robot visits the Chilean Congress

While functional and social robotics technologies may carry the potential to save lives by avoiding physical human contact and increase mental health by promoting connectedness, in a country such as Chile, initiatives such as the medicine-carrying drone can yet still be counted on the fingers of one hand. Robotics initiatives against COVID-19 are mostly pilot projects, born out of the emergency context and often led by private enterprises or start-ups, and are not based on long-term strategies. Furthermore, whenever these initiatives are reported on by the media, the focus is often on the novelty of the robot and its technology, leaving the human protagonists behind. For instance, the televised footage, showing the drone flying high above the skies of beautiful Zapallar and bringing the supplies to elderly people, does not show interviews to the people. Perhaps due to security protocols due to the pandemic, the opportunity to know how Zapallar's inhabitants feel about the drone and whether it was actually helpful to them, is lost.

As we will see next, this flying-robot drone is not the only media report in which a robot eclipses the attention from its human counterparts. "Max rather stole the show," admits Pablo Eckell, CEO of Smart Solutions Ingenieria Spa, one of the few robotics entrepreneur firms in Chile. His firm both imports and develops robots for various uses in Chile, such as customer service and health care. In October 2019, Pablo was invited by members of Chile's parliamentary Commission of Science and Technology, to give a talk about "the hurdles of developing and using robotics in the country." To this occasion, Pablo took Max, a social robot developed by his company. In the meeting, Max spoke about the possibilities that robotics brings to a commission formed by several politicians. In a voice of rather high volume, Max spoke: "we are specialists in problem solving, our only intention is to solve problems that exist daily. Robotics seeks to partake in repetitive tasks, which do not require creative intelligence from humans." Directing his words to the president of the Commission, Karim Bianchi, Max added: "Karim, the topic is complex, but I would like to close with the following reflection: for the first time you have the opportunity to dedicate yourselves to poetry instead of doing repetitive, ungratifying tasks. Don't see it as bad that all manual works are getting automatized. In truth, you will be okay as humans, because you are the ones that build the machines, you are the masters," told Max to a grinning deputy Bianchi (Miguel Mellado, 2019, Oct. 2). These words do a good job in summarizing what Eckell himself thinks about robotics: "human beings were not made for work, at least not how we think about jobs. We are made for evolution, for developing a greater conscience, this requires hard work. But we are not here to work endlessly for repetitive tasks, this a robot could do better," he tells. Apparently, Max also had a time for self-reflection while at Congress: "resign," he told another politician, "we the machines can do your job better," a sentence which sparked the laughter of several politicians and Congress staff surrounding him (Cooperativa, October 3, 2019). This sentence, which quickly got the attention and went viral on Twitter, rather took the attention away from Eckell's message. His own reflections about the hurdles and difficulties of developing and using robotics in Chile are less enthusiastic than those of Max. "Companies here do not have a long-term view," he tells. "Often, the bigger companies that we deal with have a vision of one-year terms, which is still a short-term view," he argues. Companies here sometimes take eight months to make a decision of which robot they want and what it should do, and then the year is almost over. For this reason, he thinks enterprises need to be more decisive. "Contrary to what you may believe, companies do not always know what they want, wasting enormous time and resources," he states. Regarding governmental policies, Eckell tells that he would like to see governmental initiatives to fund the provision of robotics services, instead of prioritizing funds to create robots. "Creating robots takes a lot of time, I can better import robots or robot parts, create a robot and provide my own services," he reflects.

Chilean engineer Rodrigo Quevedo, Chief Executive Officer of Robotics lab Scl, one of Chile's few manufacterers of robotic prototypes and the country's first technological hub, coincides in that governmental policies could improve. When asked what the government could do to incentivize the development of robotics in the country, he argues "robotics forces you to be at cutting edge level, but in Chile there is a technological gap. This gap causes that people don't realise about robotic services that exist or that are possible. When I go and show some of my projects to the government, they find it hard to believe that some robotic services exist." He refers to Over Mind, a system that allows people of reduced mobility to move their wheelchair via brain wave sensors. "Then [the government] questioned how many people would actually need to use this service. I told them that we are lucky that only a few. But for this reason, I go slower in certain topics, so that people can actually understand. But, beyond the government, it is the people and institutions' role to bring this technology and knowledge to the public. If we want to compete, we need to work twice as hard. I work two shifts daily. If you have a good project in Russia, you get two million dollars, in Chile with luck, you get 10 million pesos, so we are competing against that. We need to be quick and inventive to be on that level."

Eva the nurse will see you now

"The nurses here are very happy, because robots do avoid the risk of infection," Francisco Dentali, director of Verese hospital in Italy, is quoted during an interview with the national tv news in Chile (24horas, May 18, 2020). "An unexpected ally, which day by day can take on more tasks and cheers up both the medical personnel and the patients. Technology at the service of health," the voice of a Chilean journalist is heard off-screen. The images show Eva, a white robot of about 1.30-meter in height, standing on a pedestal with wheels, with a touchscreen located in the chest, arms, and a screen-face showing two cute eyes, circulating in a hospital setting. "Chile is also taking on this trend, and in our public health system: at Padre Hurtado hospital. Eva is the first pilot of a robot assistant to enter a hospital in the Coronavirus context," announces the journalist with a proud voice. Padre Hurtado, where this hospital is located, is one of Chile's poorest districts at the southwest of Chile's capital, Santiago. (See figure 10.1).

"The idea to incorporate a robot is for supporting and complementing the work of our different health teams" – the televised images now show the bust of doctor Hernán Bustamante from Padre Hurtado hospital, seen on Eva's chest, as if it were a teleconference call. Eva's face, above this screen, continues to show two friendly-staring eyes. Doctor Bustamante continues: "also, it is due to the necessity to connect with patients, because unfortunately, not every health team is available all the time."

Eva, a pioneering initiative in Chile and operational only since May 15th of this year, was facilitated thanks to an alliance between PricewaterhouseCoopers, the Faculty of Medicine at Clínica Alemana, Universidad del Desarrollo, one of Chile's private universities, and Robotics Lab. "Eva has the latest generation technology, it has sensors in its body, an automatic rechargeable battery lasting at least eight hours, a camera for visualizing and recognizing patients, a microphone, connection to the Internet, movements and 360 degree turns, besides a system of autonomic and intelligent movements, useful to perform rounds and learn about the premises to which it should adapt to," explains Rodrigo Palacios, senior

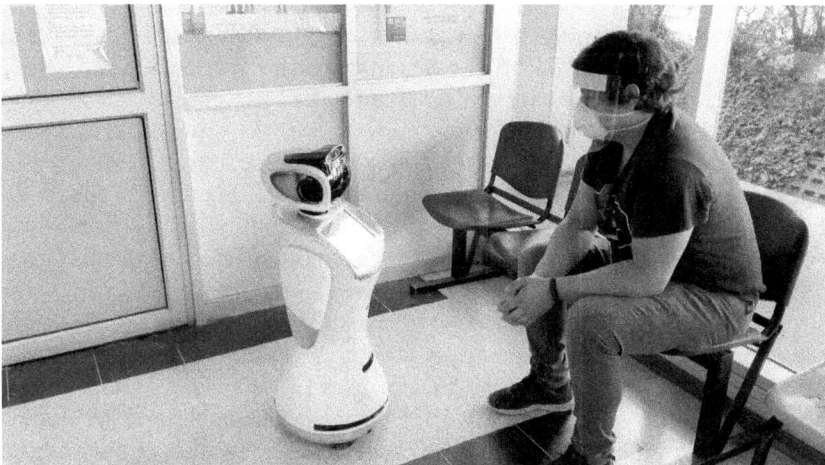

Figure 10.1 'Eva the nurse will see you now.'

manager of digital transformation at PwC Chile (Chillanonline, May 19, 2020).

Eva is one of many robot models that can help during the pandemic as an effective way to connect patients with medical staff and their families through telepresence. Telepresence refers to a person's perception that another person is "there" inside the media (Schroeder, 2002). Telepresence gives both communication partners a feeling of immersion in the communication (Witmer & Singer, 1994) and can decrease perceptions of social distance, a form of psychological distance (Trope & Liberman, 2010). Patients with COVID-19 are forced into physical and social isolation due to risk of contagion, which can result in negative cardiovascular and mental health outcomes and in increased mortality risk (Leigh-Hunt et al, 2017).

Telepresence robots provide two-way audio and video communication and can be stationary or driven into patients' rooms remotely (Nodehealth, May 19, 2020). For patients who are experiencing social isolation during the pandemic, telepresence robots can be particularly effective at increasing social connectedness, especially among the elderly (Edelman et al, 2020). The next televised image shows Eva communicating a patient to family members. "Eva has a crucial capacity to end the loneliness of patients, they now shouldn't have to talk to their families through glass or through a tablet, Eva could be at their side to keep communication as nearest as possible," adds the journalist offscreen. One great advantage of telepresence robots according to Edelman and colleagues (2020) is that in real face to face contact during COVID-19, faces can be hidden by masks or other protective equipment, while using a telepresence robot these masks are unnecessary, making communication seem more personal and warm, due to the increased facial expressions that can be conveyed.

Automated robots such as Eva can further quantify and track the healthcare workers' interactions with a patient, information which can be used for further diagnostic and or assessment purposes (Tavakoli, Carriere, & Torabi, 2020). This is the case of RAMP, a robot of preventive medical assistance which can measure the temperature, pulse, blood pressure, and lung capacity by means of a spirometry. "We have a succesful pilot running in one community health center in Macul, southeast of Santiago," tells Pablo Eckell, CEO of Smart Solutions Ingenieria Spa. "In general, people love the robots, they get closer to them and want to interact," he remarks. Other advantages are that these robots can be used to reduce workers' fatigue or stress by carrying on more strenuous repetitive tasks, such as measuring the temperature and/or by being constantly available for patients on-call. "It would clearly help us," says Javiera Macaya, staff at Padre Hurtado hospital, interviewed on screen while wearing a face shield and her doctor uniform inside the hospital premises. She states that autonomous robots such as Eva "can help educate the population about the Coronavirus but also to avoid contact, because sometimes patients keep calling on us and the robot could go to their room and see what they need," she concludes (24horas, May 18 2020). The images continue to show Eva surrounded by Padre Hurtado hospital staff wearing face masks, some of them smiling and taking pictures of Eva. "Eva has really paved the way for robots in Chile," tells Rodrigo Palacios, head of Digital Transformation and Open Innovation at PwC Chile. "Before this story broke in the media, companies or hospitals were somewhat reluctant to work with robots. Now we have companies asking us when we can bring her in."

Apart from Eva, other functional robots have been put to work against COVID-19 in other areas of the world: for instance, self-driving ultraviolet disinfection autonomous robots, since UV light has been shown to kill pathogens, have been used in the Chinese city of Wuhan, where the Coronavirus is believed to have originated. Other video-conferencing robots have been used in Shenzhen hospital in China and in Tunisia, such as the ENOVA robot (The Star, May 3, 2020). In Belgium, Zorabot, a telepresence robot, has been used to help the elderly connect with their loved ones (Reuters, March 16, 2020). Although telepresence robots are increasingly used among the pandemic, the most disregarded use of robots in Chile during the pandemic has been the social aspect. Since most robots used are either telepresence robots or functional robots. However, more sophisticated kinds of robots, such as Eva and Max, possess the capacity to have interactions with humans, and humans readily engage in social interactions with robots, who have become new social communicators (Zhao, 2006). This means that social robots could potentially act as an aid to fight feelings of loneliness and self-isolation during this pandemic. However, the use of these robots for health and mental care brings a number of ethical challenges, which will be broadly revised below.

Ethical challenges of social robots in health care

Although this chapter does not pretend to fully cover all ethical aspects concerning the use of robots in health care, it will refer to main challenges, since some very interesting aftermaths of using socially assistive robots, even before the pandemic, are the concerns regarding ethical challenges behind their use. The report by the European Parliamentary Research Service (EPRS, April 2020) highlights some ethical challenges posed by using robots during this pandemic, conceptualizing these depending on the type of robot: functional or social. For functional robots, such as UV disinfection robots, telepresence robots, or temperature-checking robots, challenges "relate to safety, radiation-related health effects and effectiveness concerns as most UV (ultra-violet) robots have only recently been deployed" (*p.* 20). Indeed, a malfunctioning UV robot could break havoc in environments which are already sensitive and perhaps lacking extra technical care such as an intensive care unit. An extra concern is that robots that perform routine checkups may mean replacing or lay off of health-care workers, the report adds, although the capabilities of these robots at present, may make this fact unlikely, given that these robots are only meant to be used for routine tasks.

In the case of social robots, or those robots that are able to interact with humans in a way that resembles human relationships (Broadbent, 2017), the EPRS report questions whether it is ethical that people could be deceived into having interactions with an entity which may appear to be and act, similar to a person. Sparrow and Sparrow (2006) posits that making people believe that a robot is something it is not, such as a real person, involves deceit and, by extension, puts the human dignity of the cared for person in question. Referring to this ethical question, Wachsmuth (2018) quotes two students discussing this issue, one student replying "would (being deceived) be so bad? We humans like to be deceived, for instance in an exciting movie." Further, regarding the issue of dignity, Wachsmuth (2018) quotes several opinions of elderly people, explaining how undignified they would feel by becoming a "burden" to a real person, perhaps a family member, or by being mortified and embarrassed to have a real nurse undress them, preferring that a robot would do this same job.

Another point of question, according to the EPRS report, is the emotional involvement and possible dependency created by the interaction with a social robot. However, these two factors, involvement and dependency, can also be common in human relations. Whereas a robot may seem like they display emotions and we might not be sure they feel them, the same can be the case with human beings, in the end, as every emotion that we witness in others, we truly cannot ontologically feel or assess in the exact same way. "It would be unclear whether a robot truly experiences or just simulates emotions. Yet the same seems to be the case in our encounters with human beings," (Wachsmuth, 2018, *p.* 2). Another point of consideration are the effects of social robots among a group of humans, going beyond one on one communication, for instance in nursing homes. A study by Wada & Shibata (2006), which introduced Paro, a cute social robot in the form of a seal, in a nursing home for one month, showed to improve the relationships between the residents themselves, whilst also decreasing their stress levels.

The EPRS report ends with some anticipatory policy-making recommendations: "efforts must be made to ensure that in the vast application of robots, their motions are predictable and are aligned with values such as transparency, accountability, explicability, auditability and traceability, and neutrality or fairness" (EPRS, April 2020, *p.* 20). The report further advises to introduce an ethical governance scheme for robotics "irrespective of whether applications are based on the capacities of artificial intelligence" (*p.* 20). Among others, it suggests that these schemes are based on the supervised autonomy of robots, fairness and privacy by design. Interestingly, in the context of the pandemic, the European Commission launched an initiative to collect ideas about deployable Artificial Intelligence (AI) and robotics solutions to create a repository easily accessible to "all citizens, stakeholders and policymakers" (European Commission, 2020). Although Chile lacks such a framework and has not made a public call for robotic initiatives to fight COVID-19, its forthcoming artificial intelligence policy does have a relevant component of "listening to the public," as it will be shown next.

Chile's "bumpy" road towards a national artificial intelligence policy

"Artificial intelligence is good news. It means that we will have new tools to face huge challenges," Chilean President Sebastián Piñera addresses a large table of experts in Chile's presidential La Moneda palace in September 2019. "We want to position ourselves at cutting-edge level in incorporating artificial intelligence, which will transform the lives of our fellow citizens. Citizens have the right to know what artificial intelligence is and how it will affect us," remarked Piñera (Prensa presidencia, Sept 12, 2019). With these words, the president paved the way for one important goal of his second presidential term: that of giving Chile its first artificial intelligence (AI) policy, a process set to hear "the voice of the people." A few days later, on August 23rd, Piñera commissioned to the Minister of Science, Andrés Couve, the drafting of a work plan to create Chile's first national AI policy. This policy "seeks to empower people in the use and development of AI tools, and make citizens participate in the debate about the legal, ethical, social and economic consequences of AI" (Ministry of Science, Technology, Knowledge, and Innovation, 2020). Since Piñera appointed artificial intelligence as a key technology and industry to develop, the Ministry of Science, Technology, Knowledge, and Innovation convened two special committees: one conformed by civil society, academics and industry experts, and the second, an interministry cabinet, led by the Ministry of Science[1]. The task of this latter cabinet is to carry out an action plan within several governmental branches, for instance that the Ministry of Science provides more scholarships for doctorates in artificial intelligence.

"Chile's process to create a public policy of artificial intelligence is unique in the world," tells José Guridi, advisor to Chile's recently created Ministry of Science, Technology, Innovation, and Technology. He explains that it is unique because the Ministry created a participatory process with different working groups, which any citizen can propose through the Ministry's website. The knowledge and conclusions gathered from these working groups will be considered in the creation of the country's first AI policy, tells Guridi. Apart from these work groups, experts, and interministry committees, extra seminars were planned to gather the vision in of Chile's regions.

"Apart from the webinars, we previously studied the AI policies of different countries," tells Guridi. In internal working groups, "we identified 27 countries that were advancing towards an AI policy. In Latin America, we found that only Mexico already had a policy, and at that moment, Argentina, Uruguay, Colombia and Brazil were in the process of creating one. Together with the Ministry General Secretariat of the Presidency

(Ministerio Secretaría General de la Presidencia), we found ourselves to be coincidentally working in the same issue, and together we revised about 12 publically available country policies and strategies in more detail," tells the advisor. From this analysis, the team defined three main areas that the Chilean AI policy should address (Equipo Futuro, 2020): 1) empowering factors; 2) development of AI and its applications; and 3) ethics, regulatory aspects, and social and economic impacts. The empowering factors consider three sub-factors: 1a) data (sources, standards, procedures); 1b) human capital, including basic, middle school, technical, bachelor, and postgraduate education and 1c) technological infrastructure, such as optical fiber, data centers, and 5G. Point 2, the development of AI and its applications, concentrates on basic and applied research in AI, and the development and demands of solutions. Lastly, point 3 refers to ethics, regulatory aspects, and the social and economic impacts of AI, apart from the opportunities that come from the good use of this technology. Some examples of issues to be dealt with in this section are the effects of AI to people's privacy, the environment, labor, gender gap, justice, and democracy, among others. The development and use of robotic initiatives would fall indirectly in point 3.

> A bill to regulate AI is not yet in the horizon, but it could happen, we leave the door open to it," adds Guridi. 'One key aspect that we are considering is that this policy should have continuity in the next presidential terms, so become a State rather than a governmental policy, which would be valid at least ten years from now.' When asked about the reasons for this, Guridi explains that technologies advance rapidly. Perhaps in five more years, we may be discussing very different technologies. Therefore, we need to ensure that this policy can run through other governments, which is complex, because in many areas we still lack baseline information that can guide our decisions. For instance, small agro-entrepreneurs in Chile have trouble conceptualizing what big data is when we ask them, therefore in some cases we have limited knowledge of how far technologies are being adopted in the country.

Although its process may be unique in the world, the road to this policy has also been unusually "bumpy" due to two main events: first, Chile's social unrest in October 2019, which saw violent clashes between police and protesters, who claimed the unrest was the "tip of the iceberg" from 30 years of severe income inequality and resulted in people injured, deaths, looting in supermarkets, and general riots in the streets (BBC, October 21, 2019; Johanson, Oct 29, 2019; Ledur & Levine, Nov 21, 2019). As a result, from this unrest, the working groups to develop this AI policy, until then carried out mostly via face to face meetings, moved to the online domain via video-chat applications. "The social unrest gave us an additional incentive to organize open online meetings to discuss this policy, which we organized from February 2020 on," tells advisor Guridi's office.

The second "bump in the road" has been the Coronavirus pandemic itself, which broke in Chile a little later than in Europe and the United States by mid-March, 2020. On March 18th, 2020, the virus reached stage 4, meaning that it started to spread in the community (Ministry of Foreign Affairs, March 19, 2020). As a result, the Chilean government decided to close all land, sea, and air borders. Soon after, quarantine periods of several weeks have been announced in the Chilean capital and main cities (Santiago Times, May 31, 2020), with the death toll continuing to rise at the time of writing, delaying the work on the policy.

Amid these events, the solution to continue to work on the policy was found in webinars, which have been organized by the Ministry of Science since May this year and are streamed on YouTube every week. These webinars seek to reach and include relevant actors from civil society, academics, experts, developers, and entrepreneurs, among others. The topics of these webinars revolve around artificial intelligence applications and services (Ministerio de Ciencia, May 10, 2020). Further, in February 2020, the Ministry launched an open call to self-convened work groups which will work through August, accessible at the Ministry's website, explained Guridi. "This process of public consultation is expected to continue until the second semester, where the president is expected to officially launch the policy," he remarked.

A public-private working group reflects about robotics through YouTube

In the three main pillars of the governments' future AI policy, robotics, and its applications (whether it refers to functional or social robots), do not figure prominently at the time of writing. In spite of this, other instances have taken the role to reflect about robotics and public policy in Chile. This is the case of the public-private working groups on robotics, organized by the Chilean-Japanese interparliamentary group of the Asian Pacific, together with a local university (Universidad de O'Higgins) and the Asian-Pacific Program of the Chilean National Library of Congress (BCN, for its Spanish acronym). Due to the pandemic, these talks have also been taken to YouTube in the form of webinars and have received attention from the local community whom, quarantined from inside their homes, are able to follow and contribute to the debates.

"Robotics is not only a mere technology, robotics also has an ethical aspect," reflects the president of the group, deputy Issa Kort, when opening the sixth session of a working group session entitled: "Robotics at the service of the pandemic: ethical challenges" (BCN, May 29, 2020). "At the end of the day, ethics belongs to one individual. What is special about ethics is that if we add up all of these individuals together, we have a collective issue," he adds. At the end of the talk, when specially consulted about the role of robotics in the Chilean national AI policy, Kort responds:

the situation of robotics in Chile is still in its infancy, this is the objective behind this panel. This panel will not resolve the situation, its objective is to analyze it and from there, make proposals. Our system is still very reliant on the president; therefore, initiatives are largely based on what happens in our government. We need to review which tools are present in our public policy, how to use them, and as well, define that robots won't replace people. This is key. For instance, when a robot that provides cleansing alcohol is seen as a replacement for people, this is the type of situation that we must be discussing. Unfortunately, robotics is not a priority within politics, this is a more emergent topic. Finally, *before* public policy is defined, I think that this debate has to reach the general public. This is not a topic to decide behind closed doors, first we need to listen to the public, academics, and experts. This is the objective behind this work group, he remarks

Blanca Borquez, a researcher at BCN adds: "our national policy of artificial intelligence is in progress. A team within the Ministry of Science, Knowledge, Innovation and Technology is working to put all the pieces of the puzzle together. The call for participation to form working groups is open."

Expert Rodrigo Verschae, PhD in engineering sciences and an academic at Universidad O'Higgins, and one of the experts in Chile's committee to develop its AI policy, remarks in closing: "from my understanding, this AI policy is still developing, but it is a rather broad call. Robotics has a rather superficial role in it. We still need to find out what the results will be. You can also participate in it. There are other research areas that have policies and specific funds. For instance, we all know that astronomy is very important in Chile and there are specific funds for astronomy. There are funds being given to specific areas and in the future, robotics could be one of them."

Note

1. The Committee is composed of 11 of Chile's ministries, among them: Education ministry, Foreign affairs ministry, National Defense ministry, apart from Chile's National Agency of Research and Development (ANID), Chile's Production Development Corporation (CORFO), and Chile's National Service of Training and Employment (Equipo Futuro, 2020).

References

24horas (2020, May 18). Así es el plan piloto para el uso de robot para combate del COVID en hospitales públicos. Retrieved from: https://www.youtube.com/watch?v=q-MGcbnBfZM.

24horas (2020, April 16). Zapallar usa dron para entregar medicamentos y mascarillas. Retrieved from: https://www.24horas.cl/coronavirus/zapallar-usa-dron-para-entregar-medicamentos-y-mascarillas-4104792.

BBC (2019, October 21). Chile protests: Cost of living protests take deadly toll. Retrieved from: https://www.bbc.com/news/world-latin-america-50119649.

BCN, Biblioteca Congreso Nacional (2020, May 29). La robótica al servicio de la pandemia: desafíos éticos [You Tube video]. Retrieved from: https://www.youtube.com/watch?v=jLKvBy5s1AI

Broadbent, E. (2017). Interactions with robots: The truths we reveal about ourselves. *Annual review of psychology*, *68*, 627–652. doi: 10.1146/annurev-psych-010416-043958.

Chillanonline (2020, May 19). Pwc lo trae a Chile: Novedoso robot se suma al combate contra el Covid-19. Retrieved from: http://www.chillanonline.cl/V5/pwc-lo-trae-a-chile-novedoso-robot-se-suma-al-combate-contra-el-covid-19/.

Cooperativa (2020, April 17). Dron salvavidas de Zapallar ahora entregará medicamentos y mascarillas. Retrieved from: https://www.cooperativa.cl/noticias/sociedad/salud/coronavirus/dron-salvavidas-de-zapallar-ahora-entregara-medicamentos-y-mascarillas/2020-04-17/103350.html.

Cooperativa (2019, October 3). Robot que visitó el Congreso embromó a Auth: "Renuncia Pepe... podemos hacer tu pega". Retrieved from https://www.cooperativa.cl/noticias/pais/politica/robot-que-visito-el-congreso-embromo-a-auth-renuncia-pepe-podemos/2019-10-03/194307.html.

Edelman, L. S., McConnell, E. S., Kennerly, S. M., Alderden, J., Horn, S. D., & Yap, T. L. (2020). Mitigating the Effects of a Pandemic: Facilitating Improved Nursing Home Care Delivery Through Technology. *JMIR aging*, *3*(1), e20110.

Equipo Futuro (2020). Política Nacional de Inteligencia Artificial. Internal report from the Ministry of Science, Technology, Knowledge and Innovation: unpublished.

European Commission (2020). AI and robotics solutions for the COVID19 crisis. Retrieved from: https://ec.europa.eu/eusurvey/runner/15e8809f-a702-fff6-b68d-9b69f98cdc2b.

European Parliamentary Research Service (2020, April). Ten technologies to fight coronavirus. Retrieved from: https://www.europarl.europa.eu/thinktank/en/document.html?reference=EPRS_IDA(2020)641543.

Johanson, M. (2019, October 29). How a $0.04 metro fare price hike sparked massive unrest in Chile. Retrieved from https://www.vox.com/world/2019/10/29/20938402/santiago-chile-protests-2019-riots-metro-fare-pinera.

Ledur & Levine (2019, November 21). 'Chile woke up': Subway fare hike sparks protests over social inequality, setting off an intense wave of unrest that has lead to at least 23 deaths. Retrieved from: https://graphics.reuters.com/CHILE-PROTESTS/0100B32527X/index.html.

Leigh-Hunt, N., Bagguley, D., Bash, K., Turner, V., Turnbull, S., Valtorta, N., & Caan, W. (2017). An overview of systematic reviews on the public health consequences of social isolation and loneliness. *Public Health*, *152*, 157–171.

Mellado, Miguel. (2019, October 2). Recibimos la visita del presidente de Smart Solution Ingenieria SPA, Pablo Adolfo Eckell, quien llegó acompañado de Max, el robot creado por su empresa, y quienes nos cuentan sobre su asombroso proyecto! [Tweet]. https://twitter.com/melladosuazo.

Ministry of Foreign Affairs (2020, March 19). Information COVID19. Retrieved from https://minrel.gob.cl/covid-19-coronavirus-outbreak-recommendations/minrel/2020-03-19/115108.html.

Ministry of Science, Technology, Knowledge and Innovation (2020). Proceso de participación para contribuir con la Política Nacional de Inteligencia Artificial. Retrieved from: http://www.minciencia.gob.cl/politicaIA.

Ministerio de Ciencia (2020, May 10). Ministerio de Ciencia impulsa webinars con expertos para avanzar en la política de inteligencia artificial. Retrieved from: http://www.minciencia.gob.cl/noticias/ministerio-de-ciencia-impulsa-webi-nars-con-expertos-para-avanzar-en-la-politica-de-inteligencia-artificial.

Nodehealth (2020, May 19). Digital health tools for the emergency department before during and after Covid19. Retrieved from https://nodehealth.org/2020/05/19/digital-health-tools-for-the-emergency-department-before-during-and-after-covid-19/.

Núñez-Torrón (2020, April 23). En este pueblo chileno los drones entregan mascaril-las y medicinas a los mayores aislados. Retrieved from: https://www.ticbeat.com/tecnologias/drones-chie-comida-medicina-mayores-aislados/

Prensa presidencia (2019, September 12). Presidente Piñera se reúne con senadores y científicos para tratar política de inteligencia artificial. Retrieved from site: https://prensa.presidencia.cl/fotonoticia.aspx?id=101529.

Reuters (2020, March 16). Belgian video-calling robots to keep elderly connected dur-ingcoronavirus.Retrievedfrom:https://www.reuters.com/article/us-health-coronavirus-belgium-robots/belgian-video-calling-robots-to-keep-elderly-connected-during-coronavirus-idUSKBN21339G.

Santiago Times (2020, May 31). Chile exceeds 1,000 deaths, Coronavirus cases close to 100,000. Retrieved from: https://santiagotimes.cl/en/2020/05/31/chile-exceeds-1000-deaths-coronavirus-cases-close-to-100000/.

Schroeder, R. (2002). Social interaction in virtual environments: Key issues, com-mon themes, and a framework for research. In R. Schroeder (Ed.), *The social life of avatars: Presence and interaction in shared virtual environments.* London: Springer-Verlag.

Sparrow, R., & Sparrow, L. (2006). In the hands of machines? The future of aged care. *Minds and Machines, 16*(2), 141–161. doi: 10.1007/s11023-006-9030-6.

Tavakoli, M., Carriere, J., & Torabi, A. (2020). Robotics, Smart Wearable Technologies, and Autonomous Intelligent Systems for Healthcare During the COVID-19 Pandemic: An Analysis of the State of the Art and Future Vision. *Advanced Intelligent Systems*, 2000071.doi: 10.1002/aisy.202000071.

The Star (2020, May 3). Robot helps Tunisia medics avoid infection from virus patients. Retrieved from: https://www.thestar.com.my/tech/tech-news/2020/05/03/robot-helps-tunisia-medics-avoid-infection-from-virus-patients#cxrecs_s.

Y. Trope, N. Liberman. (2010) Construal-level theory of psychological distance. *Psychological Review*, 117(2), doi: 10.1037/a0018963.

Wachsmuth, I. (2018). Robots like me: Challenges and ethical issues in aged care. *Frontiers in psychology*, 9, 432.

Wada, K., & Shibata, T. (2009). Social Effects of Robot Therapy in a Care House– Change of Social Network of the Residents for One Year–. *Journal of advanced computational intelligence and intelligent informatics*, 13(4), 386–392. doi: 10.20965/jaciii.2009.p0386.

Witmer, B. G., & Singer, M. J. (1994). *Measuring immersion in virtual environments.* Alexandria, Virginia, U.S. Army Research Institute for the Behavioral and Social Science.

Zhao, S. (2006). Humanoid social robots as a medium of communication. *New Media & Society*, 8(3), 401–419. doi: 10.1177/1461444806061951.

11 Intellectual property rights and social media policies for user-generated content

Some lessons from Mexico

Rosa María Alonzo González

Introduction

Social media platforms as Instagram, YouTube or TikTok are exceptional spaces in where the possibilities of socializing, the desire for social approval, and the opportunity to monetize content seduces users, mainly the children and youth which are constantly consuming and sharing information using smart phones and tablets. In Latin America, there is a growing market for the consumption of content shared on social media and streaming services. According to some estimates in 2019, about 289 million people were regular consumers of streaming videos. This number, however, will rise in 2023 to nearly 50% of the population (Ceurvels, 2019). Latin Americans are enthusiastic fans of online series and movies offered by Netflix or HBO, and avid consumers of social media content generated and published on YouTube by users known as *YouTubers*. Many Latin American YouTubers such as *Yo soy German*/I am German or *Yuya* have millions of subscribers, and their annual revenues can easily surpass millions of US dollars. *Luisito Comunica,* a Mexican vlogger in YouTube, for example, got four million dollars in gains in 2018 (Redacción El Universal, January 3, 2018). This growing market is sparked by two key facts: (1) a rising market related with the consumption of user-generated content. Most of these users (vlogger, influencers, or YouTubers) are not necessarily related with traditional media corporations as television networks. On the contrary, some of them started as amateurs sharing their content for ludic purposes only; (2) Latin American children and young ones are devoted consumers of user-generated content (Bailey et al., 2018). Youtubers as *Yuya*, whose content focused mainly on teenagers, has nearly 24.2 million subscribers. However, this rising market is not immune to legal disputes, particularly considering the intellectual property rights of the material shared inside these platforms.

In 2014, many well-known Mexican social media influencers like *Werevertumorro* were forced to temporarily eliminate their content from YouTube due to intellectual property issues. This, however, was not caused by a violation of the social network's user policy. A company, entitled "W2M Enterprises," which managed a Multi-Channel Network (MCN) or Network of successful Latin American YouTubers such as *Yuya, Yo soy German* and *Werevertumorro*, appropriated a significant share of the revenues generated by these vloggers, and in the case of *Werevertumorro*, seized his own name. This case had deep legal consequences for the user-generated content market, particularly regarding the copyright of the material produced and commercialized on social media. This chapter will focus on the analysis of The Werevertumorro case, as it is widely known. It will be possible to observe in this chapter that besides the important growth of the user-generated content market in Latin America, there are important legal deficiencies that should be amended to guarantee the economic and social sustainability of the market in the forthcoming years. This is not only important for those companies and individuals who commercialize their content in platforms such as YouTube or TikTok, as in the case of YouTubers or influencers, but to all individuals and social organizations who make use of social media as forums for political and social expression as well.

Intellectual property of digital media

Traditionally, technology public policies in Latin America have focused mainly on reducing the negative effects associated with the digital divide and to promote social and economic development with the support of Internet: the lack of connectivity, deficient technological infrastructure, and low levels of digital literacy among the population. The Internet, since its arrival into Latin America in the mid-nineties of the 20th century, has been avowed by governments and civic organizations as an important path to increase governability, social participation, education, and community development. For many countries, the Internet is not just another technology but a human right. In 2013, Mexico granted the access to the Internet as a constitutional right (Ormaetxea, 2017). One year later, in 2014, Brazil approved a legislation that assures the equal access and the right of privacy for all Brazilian users (Boadle, 2014). But as the number of users and services diversifies and grows, the request for updating and improving legal frameworks that protects user privacy, free expression, and property rights in virtual environments becomes an imperative. Intellectual property rights (IP) are particularly important regarding social media platforms in where individuals share diverse protected multimedia who are not necessarily licensed to do so, but at the same time, they are exposed to a potential stealing and misuse of the content they create and publish. Some cases not only compromise private interests but collective ones as well. For example, digital videos and photographs depicting indigenous ceremonies, vestments, and cultural artefacts are distributed on social media without the consent of the indigenous communities (Robertson, 2014). These incidents not only stand a violation of the communities' IP rights, but many times these patterns, sacred adornments, and symbols are illegally copied and employed in the manufacturing of designer clothing by international fashion companies (Rea, 2020).

In spite of the growing importance of IP laws for protecting individual and corporative content, these regulations are not new, but on the contrary, they have a long tradition in Latin America. Cerda Silva (2016) associated their origins to the national independence and the later arrangement of the first national legal codes. While the first Latin American's constitutions granted some kind of authorship rights, it was not until the Universal Copyright Convention in 1952, the subsequent acknowledgement of the Berne Convention for the Protection of Literary and Artistic Works and the Rome Convention for the Protection of Performers, Producers of Phonograms, and Broadcasting Organizations, that Latin American countries counted with a defined framework that protected intellectual property rights. But even with these conventions, treaties, and the later incorporations to local regulations, implementing traditional normativity molded to protect tangible goods as books, DVDs, and CDs to shelter digital content is not an easy task (European Commission, 2019). Every day, several platforms allow the interchange of millions of multimedia elements. The inner logic of social networks encourages the quick share and access to the information, making it almost impossible to properly track all the flow data. This exchange may contain protected works that require a specific permission from the copyright holder, but there is also material created and distributed free of charge by the users themselves. Those videos, photographs, and audio files do not unavoidably remain static inside the platforms, but they are susceptible to being distributed or even modified by other individuals. One huge barrier that limits the legal protection of the material published on social media is the high speed in where theses exchanges occur. Gargantuan social media platforms as the case of YouTube (with 1.8 billion users) are in constant movement and expansion (Gilbert, 2018). Trying to surveil the huge number of videos and commentaries shared by users all over the world and comply with national and international IP regulations is challenging. This complexity is not exclusive to the protection of copyrights or *Derechos de Autor*, as it is called across Latin America, but to all regulations related with the Internet: privacy, cybercrime, and free speech, among others.

Social media and the user-generated content

Besides, social media is widely known as a unified technology, it is diverse and assorted. Social media is a galore of applications and virtual platforms accessible through different devices as smartphones, computers, tablets, video game consoles, televisions, and so on (Miller et al., 2016). Social media enables the production, edition, publication, diffusion, and reception of virtual content. The nature of this content is varied such as the users' reasons to create and share content. From ludic to satirical, political or artistic (Hemsley et al., 2018), social media operates under the principles of Web 2.0 (O'Reilly, 2007). Here are some of their characteristics:

- These platforms back users to creat social networks which promote high participation, with a considerable number of connections and interactivity among the members.
- They offer virality and social influence.
- Social media is easy to use, allowing a wide sector of the population to create and publish content.
- They offer horizontal channels of interpersonal and social communication (from many to many), without intermediaries.
- Social media platforms have a potential global scope.
- They establish sufficient conditions for freedom of expression; however, they are expected of being subjected to governmental restrictions thought user policies.

Besides these characteristics and potential benefits, there are some critic voices that consider these spaces as potential threats to democratic values and peoples' rights. Christian Fuchs (2014), for example, point outs at least three barriers that conditions the user's approach to social media: (1) An economic antagonism amid the peoples' privacy right and the interest of media companies to obtain benefits from the content shared by individuals. In this sense, the user' private data turns out to be a currency. (2) A political antagonism between user privacy, and corporative and governmental surveillance. As people's social activities rely everyday more on these platforms, there are rising concerns about how authorities and companies could misuse data not just to apply abusive adverting campaigns but surveil and prosecute political dissidents and (3) The monitoring and censorship of public discourse online. The government agencies and corporations could eventually limit, feudalize, and colonize the online public discourses. Consequently, social media does not turn out to be free and democratic, as it might supposed to be; but instead, it will condition its services for uploading, sharing digital content, controlling the communication flow, and requesting.

Even before the eruption of the Internet and social media, people already produced and consumed content generated by professional artists for their audience: readers, radio listeners, or television viewers. This content embraced telephonic commentaries in radio stations, alternative publishing supported by enthusiasts as the case of fanzines, letters to the magazine editors, etc. (García-de-Torres, 2010). However, the arrival of Internet has drastically changed, not just how people interact with information, but how they could eventually become producers and consumers at the same time. From a certain point of view, the Internet has democratized both the information and the tools to produce and share content. Actually, digital media enables the access to a myriad number of digital contents; some of this material is generated by the users themselves and is distributed under diverse formats and modalities: text circulated on blogs or web pages, videos published on YouTube or Vimeo, images, and photographs, spread through Instagram, among others. To access this content consumers only require a profile, generally linked to an email account and an electronic device connected to the Internet.

Generally, user-generated content implies at least three main characteristics: an observable creative effort, created outside of the professional dynamics, and published in open digital format to be accessible through the Internet (OECD, 2006). Under the label of "user-generated content" is embraced all content available through social networks and online platforms, created and distributed by one or more non-professional individuals. The result can be the invention of new content or the adaptation of previous content, always freely and voluntarily (Fernández Castrillo, 2014). The proliferation of user-generated content is a decisive advance towards the democratization of information technologies in Latin America, since it encourages the growth of Internet adoption not just to access information but to produce and generate content. Latin Americans are no longer static consumers of the information distributed by transnational media corporations but distributors of contents that reflect their political and social aspirations. In addition, it establishes new cooperation mechanism to encourage creativity and innovation.

But as the social media platforms like YouTube or Instagram consolidated as a popular source of entertainment in Latin America, many amateur creators started getting huge revenues for their content. They left amateurism and moved to the self-professionalization of the activity. User-generated content entered a commercial dynamic based on the number of "views" and "followers" that a particular social media profile received. Media and marketing corporations understood this shift as an opportunity to approach to a younger generation with new habits of consumption (Tabares, 2019). Actually, many famous YouTubers operate as a company with content editors, community managers, and marketing offices. But, as it will be possible to observe in the following section, international and local legal frame works stayed far behind from this commercial shift.

W2W enterprises vs *Werevertumorow* case

In 2006, Time Magazine presented in its cover *"You"* as the person of the year to refer to people who spend part of their time creating content for the Internet. This magazine speaks about a generation of content creators who take advantage of the potential of The Web 2.0 (O'Reilly, 2007). Today millions of people can express and communicate freely, connect, and share content, but all this is possible through social media and other digital platforms that could easily be used to upload and spread the user-generated content around the world. Each platform specializes in some particular dynamics and service. Social media allows users to spread user-generated content but, by doing that, they request information and the acceptance of their policies. These policies tend to adopt measures to protect the copyrights and intellectual property. However, they are inclined to benefit business brands, trademarks, or content with copyrights. In addition, social media policies have some legal loopholes that create risks for the protection of intellectual property of user-generated content, which creates conflicting situations mainly when these contents are susceptible of being monetized or even profitable.

YouTube is the second most used social platform in the world (Kemp, 2020). In Latin American countries such as Mexico, this social media is among the three most popular platforms (IAB, 2019; Kemp 2020). YouTube belongs to the Google company, a giant media corporation specialized in Internet-related products and services. YouTube is considered as one of the top social networks that introduced deep changes in the way people interact with content (Pardo, 2016); and settled new forms of reception and production (Bañuelos, 2009). Since 2007, this platform has a YouTube Partner Program that allows its users to monetize the uploading of videos and channels, as long as this program is available in the user's country (YouTubeTeam, December 10, 2007). This program is accessible in the following Latin American countries: Argentina, Bolivia, Brazil, Chile, Colombia, Costa Rica, Dominican Republic, Ecuador, El Salvador, Guatemala, Honduras, Jamaica, Mexico, Nicaragua, Panama, Paraguay, Peru, Puerto Rico, Uruguay, and Venezuela. This platform has a YouTube Creator Academy that, since 2014, has offered its content in Spanish to support the professionalization of Spanish-speaking creators. (Leicht, W., October 14, 2014). Doing this, the platform ensures to encourage its users to know the platform's policies as content creators and is useful to keep users informed about changes in user policies and how to monetize their video channels (Mohan, N., January 16, 2018). The following passages will focus on the analysis of copyright cases associated with user-generated content on YouTube. Some of them were huge frauds caused by the loss or transfer of the copyrights of these contents. YouTube, as a digital platform, has been kept out of such issue. This situation raises questionings about how much liability the digital platform has regarding conflicting content.

Werevertumorro is the username of a Mexican influencer. He, along with other YouTubers, were involved in the most famous case of copyright concerning user-generated content in Latin America. The Mexican press (*El Universal, Excelsior, Publimetro, Unocero, Milenio, El informador*) and entertainment magazines (*Quién, Entepreneur*) gave a wide coverage to this case, and it was plentifully documented in diverse videos published on YouTube. Currently, his case has placed a reference in Mexican schoolbooks (DannielRo, January 30, 2020).

Gabriel Montiel, better known as *Werevertumorro* or Gabo, started uploading videos on the YouTube platform in 2007, and at the age of 18, he became famous for his comedic videos characterized by his creativity and distinct sense of humor. These videos began as a recreational hobby with almost 5000 views, but his daily videos about the 2010 FIFA World Cup, South Africa, reached 300,000 views (Montiel, G., January 29, 2017a). Since 2010, he began improving the structure and production of his material. He incorporated other people to support him in the production of the videos. This was the beginning to form a team of video creators, known as *Werevertumorro Crew* – integrated by Gabriel Montiel – *Luisitorey, Fedelobo, Criss Cross, Isra, Wereverwero, Escorpión Dorado/Alex, and Fex*, whom collectively or individually created videos for many channels on YouTube. Eventually many of them have become popular YouTubers as well.

Amid 2010 and 2011, Gabriel Montiel generated videos for YouTube and, at the same time, created a company called *W2W Enterprises S.A. de C.V* with a business agent, Javier Talán and other investors. This company created a Multi-Channel Network (MCN) or Network on YouTube called *Werevertumorro* in which the channels of other famous YouTubers in Mexico were integrated (Redacción El Universal, April 1, 2014). At that moment, Gabriel Montiel was the most popular video creator of YouTube in Mexico (Gómez, B, April 3, 2014, Montiel, G., April 3, 2014b), so his presence as a partner of the company *W2W Enterprise* was used to lure more YouTubers under the same MCN (Montiel, G., January 29, 2017a) . The company *W2W Enterprise* and the *MCN Werevertumorro* worked under a contract that included managing the monetization of all the YouTubers' videos and the channels through a single MCN. This contract allowed *W2M Enterprises* to manage these YouTuber's performance with brands and products. These influencers started gaining a greater public outside of the YouTube platform. The *Werevertumorro Crew*, also known as *W2MCrew*, began to obtain presentations on radio and television programs, brand representations, and other types of marked promotional products that were advertised through the videos and channels on YouTube. Likewise, they created a live action show called the *Werevertumorro show*, and it had presentations throughout all the country (Notimex, July 31, 2011).

Although all of these YouTubers signed a contract in 2011, it was until 2014 that the *Werevertumorro Crew* began to notice discrepancies with the revenues administered by their manager Javier Talán (Redacción Excélsior, March 28, 2014), and there were even no reported earnings to these influencers for their activities and videos (Montiel, G., January 29, 2017c). This situation coincided with the request to sign a new contract for the *W2MCrew* (Montiel, G., May 7, 2014c), this action caused suspicion among the YouTubers. Other sources point out that the fraud was perceived when the company *W2M Enterprise* tried to collaborate with a foreign company (Murata, G., April 3, 2014; Redacción Quién.com, 2014, April 3). This foreign venture company wanted to renew the contracts with the YouTubers who were part of the *MCN Werevertumorro*, and upon reviewing the financial situation, the mismanagement came to light and some demands began to arrive towards Gabriel Montiel (Montiel, G., April 3, 2014b, Murata, G., April 3, 2014; Redacción Quién.com, April 3, 2014). Gabriel Montiel did not know all of the management of the *W2M Enterprise* company of which he was a partner (Montiel, G., March 21, 2014a; Montiel, G., May 9, 2019; Montiel, G., January 25, 2020).

In March 2014, the *Werevertumorro Crew* began to remove all videos and unlink all their channels to the *MCN Werevertumorro* on YouTube. This action deviated from an outcome on a legal dispute about the intellectual property rights of their videos and channels and the copyright of the werevertumorro branch which involved the loss of the monetizing in favor of their manager, who had the economic rights of the videos (Montiel, G., May 9, 2019). To explain the situation, Gabriel left only two videos on his channel, "Adios Werevertumorro" and another called "Regresaremos ...?" (Redacción Excélsior, March 28, 2014). "Adios Werevertumorro" reached 9,398,605 views. In this video Gabriel Montiel and the rest of Werevertumorro Crew briefly advised their subscribers that they would stop uploading videos to their various channels on YouTube, an action that had been carried out in an interrupted and growing way for seven years (Montiel, G., March 21, 2014a).

Gabriel Montiel gave up 80% of his income in his contract with Javier Talán (Redacción Publimetro, February 1, 2017). This situation affected him in three ways: (1) first as a content creator by the temporally loss of his videos and channels on YouTube. He recovered his usernames and passwords and could continue uploading videos to YouTube, but MCN Werevertumorro charged for their monetization, (2) the possible loss of the brands integrated into their videos, such as the name of *Werevertumorro* and others generated for channels and videos, which Javier Talán was trying to legally register under his name in the Instituto Mexicano de Propiedad Intelectual (IMPI)/ Mexican Institute of Intellectual Property (Montiel, G., April 3, 2014b), and (3) for being a partner of the *W2M Enterprise*, Gabriel was receiving lawsuits from the other influencers and YouTubers who were affected by this company and its network (Montiel, G., April 3, 2014b, Murata, G., April 3, 2014). *W2M Enterprise* and *MCN Werevertumorro* affected other Mexican YouTubers like *Yuya, Caelike (now Caeli), MisaelVlog, Miranda Ibañez, Cess Leon, Mexiblogs-Jaguar-u,* and other Latin American YouTubers such as *Hola, soy Germán* from Chile and *IAMDVDX* from Colombia (Montiel, G., April 3, 2014b, Montiel, G., January 29, 2017b).

Each contract of *W2M Enterprise* and *MCN Werevertumorro* specified different conditions for each YouTuber. For example, the contract made with *Yuya* (Mariand Castrejón), states the possibility of invoicing and collecting on behalf of her (Pineda, C., April 5, 2014). Like other influencers, Yuya announced in a video that she was having problems with the MCN that administered her channel. She indicated that someone wanted to steal her work, and she would seek a legal solution to recover her channel (Castrejón, M., March 26, 2014). YouTubers from other countries besides Mexico, such as *Hola, Soy Germán* (Germán Garmendia), had supposedly signed the contract, but the laws of his country had not allowed its execution (Montiel, G., January 29, 2017b). However, Germán pointed out in his video that he did not sign the contract and thus evaded any legal conflict (Germán, G. December 9, 2016).

In 2017, after a three-year legal battle, Gabriel Montiel recovered the trademark registration of *Werevertumorrow* and the monetization of all his channels and videos in YouTube that he had lost due to the contract.

The sum of money he lost in the legal battle was between half a million dollars to 10 million dollars, but he recovered the register of his brands, the videos created with his intellectual property and economic rights, his YouTube channels, and the possibility to monetize them (Montiel, G., January 29, 2017d; Redacción Publimetro, February 1, 2017). Today Gabriel Montiel has the trademark registered in the IMPI of all those creative mark works developed for his videos such as *Werevertumorro, PelusaCaligari, Niños del campo, Adultescentes, Mexmen, Futbol por la banda, La lata*, among others. (León, A. and Pineda, C., April 9, 2014; Montiel, G., January 29, 2017c

Even though Gabriel Montiel and Mariand Castejón (Yuya) did not point the liability of YouTube as a part of this legal situation that affected them in their videos, it is a fact that they do not receive any support from the platform to recover their channels or videos. They point out that the platform kept out of the situation and demarcated, placing impartiality in the matter (Murata, G., April 3, 2014b). Likewise, Gabriel questions himself about the liability of the YouTube platform for allowing Javier Talán to register the *MCN Werevertumorro* with his username without his acceptance (Montiel, G., April 3, 2014b). According to YouTube policies, this platform respects and protects the copyright and intellectual property of the content that users share and takes care of any content that infringed the intellectual property of third parties. The penalization for copyright or intellectual property offenses in the YouTube policies could differ from the penalization of one video with its removal to the service suspension of the user that infringes copyright. However, the process is different between a video that infringes other user's generated content from videos that violate a brand, trademark, copyright, or other work legally registered. Therefore, even when YouTube policies have a process to follow up all complaints about content copyright infringement, this examination is limited, and the result is almost the same with the content removal in favor of the claimant, who presents and justifies his/her claim. In this sense, the YouTube policies do not protect a username like a brand or the content as copyright content in the videos unless they can legally prove that they are registered trademarks, or the content has some legally registration.

Conclusion

During 2014, when the Werevertumorro case was revealed in the press, there were other similar but lesser known incidents, such as the Youtuber *Braindeadly* with the MCN Machinima (Unocero Redacción, March 31, 2014). These situations continue happening in the same way nowadays, with frauds occasioned by the intermediation of the Multi-Channel Network (MCN) better known as Network (López, M., January 31, 2020). The most recent case has been exposed about the influencer Superholly about the Latin American MCN MiTú, and even though her time was short on the network and she lost little money, she observes that this network uses her presence to integrate other novel YouTubers (Superholly, January 19, 2020). Another example in Mexico is the YouTube channel *De mi rancho a tu cocina*, which has gained a large number of subscribers in a matter of one year (Redacción Excelsior., June 16, 2020). This channel has given the opportunity to a 69-year-old woman who appears in the videos to be considered one of the 100 most powerful women in Mexico by Forbes Mexico magazine (Hernández, M., June 15, 2020). The fame achieved by this woman in 2019 generated doubts about the intellectual property rights of the videos on YouTube, particularly about the legal permission to share her recipes on YouTube and if she received some revenues for the monetization of the content (Tu Cosmopolis., September 27, 2019; Gallego, J., October 4, 2019). These cases related with user-generated content in Latin America provide important lessons for the future of the market:

1 Social media platforms are services for sharing content, not to legally protect the production, so their liability is limited.
2 The intellectual property right of the user-generated content in digital platforms is recognized as a moral right, as long as there is no claim from a third party, in which case it will proceed according to the social media policies.
3 The economic rights that users get for their content could be subject to agreements out of social media policies, in that case the social media can stay impartial, so any future conflict must be resolved according to the national laws.
4 It is useful to have legal registration of the content, copyright, and trademarks before uploading a user-generated content, because there is always a possibility to gain profit out of it.

References

Bailey, A. A., Bonifield, C. M., & Arias, A. (2018). Social media use by young Latin American consumers: An exploration. *Journal of Retailing and Consumer Services*, 43, 10–19. https://doi.org/10.1016/j.jretconser.2018.02.003

Bañuelos, D. J. (n.d.). *YouTube como plataforma de la sociedad del espectáculo*. 25.

Boadle, A. (2014, April 23). Brazilian Congress passes Internet bill of rights. *Reuters*. https://www.reuters.com/article/us-internet-brazil-idUSBREA3M00Y20140423

Castrejón, M. (2014, March 26). *¿Me voy de youtube? - YUYA* [Video en youtube]. https://www.youtube.com/watch?v=7dwm75QvKTY

Cerda Silva, A. J. (2016). Evolución histórica del Derecho de Autor en América Latina. *Ius et Praxis*, 22(1), 19–58. https://doi.org/10.4067/S0718-00122016000100002

Ceurvels, M. (2019). *Latin America Digital Video 2019*. eMarketer. https://www.emarketer.com/content/latin-america-digital-video-2019

DannielRo. (2020, January 30). @*werevertumorro lo que uno se encuentra en la escuela* ☻*https://t.co/nPTLkaCPU6"/Twitter*. https://twitter.com/Christian_Dani_/status/1222986205442269184

European Commission. (2019). *Report on the protection and enforcement of intellectual property rights in third countries*. European Commission.

Fernández Castrillo, C. (2014). Prácticas transmedia en la era del prosumidor: Hacia una definición del Contenido Generado por el Usuario (CGU). *CIC. Cuadernos de Información y Comunicación, 19*, 53–67. https://doi.org/10.5209/rev_CIYC.2014.v19.43903

Fuchs, C. (2014). Retos para la democracia. Medios sociales y esfera pública. *Telos: Cuadernos de Comunicación e Innovación, 98*, 71–82.

Gallego, J. (2019, October 4). *La estafa detrás del canal de mi rancho a tu cocina* [Video en yutube]. https://www.youtube.com/watch?v=XvbNJNr1srM

García-de-Torres, E. (2010). Contenido generado por el usuario: Aproximación al estado de la cuestión. *Profesional de la Información, 19*(6), 585–594. https://doi.org/10.3145/epi.2010.nov.04

Germán, G. (2016, December 9). *El día que casi lo pierdo todo (el significado de otra vez)* [Video youtube]. https://www.youtube.com/watch?v=CtJlGBu9e3E

Gilbert, B. (2018, May 4). How many people use Youtube? Over 1.8 billion. - Business Insider. *Business Insider.* https://www.businessinsider.com/youtube-user-statistics-2018-5?r=MX&IR=T

Gómez, B. (2014, April 3). Werevertumorro te da sus claves para ser un youtubero estrella en México. *Entrepreneur, Digital.* https://www.entrepreneur.com/article/266843

Hemsley, J., Jacobson, J., Gruzd, A., & Mai, P. (2018). Social Media for Social Good or Evil: An Introduction. *Social Media + Society, 4*(3), 2056305118786719. https://doi.org/10.1177/2056305118786719

Hernández, M. (2020, June 15). Doña Ángela de "De mi Rancho a tu cocina" es una de las 100 mujeres líderes de México. *El Universal.* https://www.eluniversal.com.mx/menu/dona-angela-de-de-mi-rancho-tu-cocina-es-una-de-las-100-mujeres-lideres-de-mexico

Interactive Advertising Bureau (IAB). (2019). *Estudio de consumo de medios y dispositivos entre internautas mexicanos* (Resumen ejecutivo No. 19; Estudio de consumo de medios y dispositivos entre internautas mexicanos, p. 51). Interactive Advertising Bureau (IAB. https://www.iabmexico.com/estudios/estudio-de-consumo-de-medios-y-dispositivos-entre-internautas-mexicanos-2019/

Kemp, S. (2020). *Digital 2020 Global Digital Overview* [Resumen ejecutivo]. We are social. https://wearesocial.com/digital-2020

Leicht, W. (2014, October 14). The YouTube Creator Academy goes global. *YouTube Creator Blog.* https://youtube-creators.googleblog.com/2014/10/the-youtube-creator-academy-goes-global.html

León, A., & Pineda, C. (2014, April 9). La disputa legal que ganó Werevertumorro. *El Universal.* https://archivo.eluniversal.com.mx/espectaculos/2014/la-disputa-legal-que-gano-werevertumorro-1001966.html

López, M. (2020, January 31). Cuidado con los fraudes dirigidos a YouTubers. *Unocero: Especialistas en tecnología y estilo de vida digital.* https://www.unocero.com/entretenimiento/fraude-creadores-youtube/

Miller, D., Costa, E., Haynes, N., McDonald, T., Nicolescu, R., Sinanan, J., Spyer, J., Venkatraman, S., & Wang, X. (2016). What is social media? En *How the World Changed Social Media* (1.a ed., Vol. 1, pp. 1–8). UCL Press; JSTOR. https://www.jstor.org/stable/j.ctt1g69z35.8

Mohan, N. (2018, January 16). Additional Changes to the YouTube Partner Program (YPP) to Better Protect Creators [Intitucionale Youtube]. *YouTube Creator Blog.* https://youtube-creators.googleblog.com/2018/01/additional-changes-to-youtube-partner.html

Montiel, G. (2014a). *Adios Werevertumorro Werevertomorro* [Video en youtube]. https://www.youtube.com/watch?v=RLyY7fTDfsg&feature=youtu .be&list=UUzVIrPfZBE-XkBISBybMBLA

Montiel, G. (2014b). *Werevertumorro por siempre Werevertumorrow* [Video youtube]. https://www.youtube.com/watch?v=85gjWliVNTY

Montiel, G. (2014c, May 7). *Vida Cruel 15—Los contratos* [Video en youtube]. https://www.youtube.com/watch?v=Bn7iAnWumJA

Montiel, G. (2017a, January 29). *El día que perdí todo: El contrato (parte 1)* [Video en youtube]. https://www.youtube.com/watch?v=KS-vbDqXKQs&feature=youtu.be

Montiel, G. (2017b, January 29). *El día que perdí todo: Los youtubers (parte 2)* [Video en youtube]. https://www.youtube.com/watch?v=sPdT3aMXgYc

Montiel, G. (2017c, January 29). *El día que perdí todo: 10 millones de pesos (parte 3)* [Video en youtube]. https://www.youtube.com/watch?v=Vd0LUDlpXkI

Montiel, G. (2017d, January 29). *El día que perdí todo: Ganamos (parte 4)* [Video en youtube]. https://www.youtube.com/watch?v=zeF-4KUDeN4

Montiel, G. (2019, May 9). *W2MCrew La confesión de los hechos (Audios de exmanager jamás revelados)* [Video en youtube]. https://www.youtube.com/watch?v=JTHAEDG7z54

Montiel, G. (2020, January 25). *W2MCrew vs Patas Chuecas: Aqui te lo cuento* [Video en youtube]. https://www.youtube.com/watch?v=irJgTiRzbHw

Murata, G. (2014, April 3). Me están doblemente chingando: Werevertumorro. *Milenio.* https://www.milenio.com/espectaculos/me-estan-doblemente-chingando-werevertumorro

Notimex. (2011, July 31). "Werevertumorro show", jóvenes famosos en Youtube. *El Informador :: Noticias de Jalisco, México, Deportes & Entretenimiento.* https://www.informador.mx/Tecnologia/Werevertumorro-show-jovenes-famosos-en-Youtube-20110731-0093.html

OECD. (2006). *Participative Web and User-Created Content: Web 2.0, Wikis and Social Networking.* Organisation for Economic Co-operation and Development. https://www.oecd.org/sti/ieconomy/participativewebanduser-createdcontent-web20wikisandsocialnetworking.htm

O'Reilly, T. (2007). What Is Web 2.0: Design Patterns and Business Models for the Next Generation of Software. *International Journal of Digital Economics, 65,* 17–37.

Ormaetxea, A. (2017, octubre 5). En México, el acceso a internet es un derecho constitucional. *Expansión.* https://www.expansion.com/latinoamerica/iberoamericana-empresarial/2017/10/15/59e3612c46163fa16d8b4615.html

Pardo Abril, N. G. (2016). El discurso multimodal en Youtube. *Revista Latinoamericana de Estudios del Discurso, 8*(1), 77. https://doi.org/10.35956/v.8.n1.2008.p.77-107

Pineda, C. (2014, April 5). No están facultados para facturar por Yuya. *El Universal.*/espectaculos/2014/impreso/-8220no-estan-facultados-para-facturar-por-yuya-8221-130322.html

Rea, C. (2020, junio 12). México acusa a Carolina Herrera de plagiar textiles indígenas: ¿fue robo o inspiración? *Univision.* https://www.univision.com/noticias/america-latina/mexico-acusa-a-carolina-herrera-de-plagiar-textiles-indigenas-fue-robo-o-inspiracion

Redacción El Universal. (2014, April 1). El Werevertumorro perdió hasta el nombre en YouTube. *El Universal.*/espectaculos/2014/el-werevertumorro-perdio-hasta-el-nombre-en-youtube-999755.html

Redacción El Universal. (2018, January 3). ¿Cuanto ganan los youtubers mexicanos? *El Universal.* https://www.eluniversal.com.mx/de-ultima/cuanto-ganan-los-youtubers-mexicanos

Redacción Excélsior. (2014, March 28). Lo que ocurrió con Werevertumorro en Youtube. *Excélsior.* https://www.excelsior.com.mx/hacker/2014/03/28/951133

Redacción Excélsior. (2020, June 16). Doña Ángela... Entre las 100 mujeres más poderosas de México. *Excélsior.* https://www.excelsior.com.mx/nacional/dona-angela-entre-las-100-mujeres-mas-poderosas-de-mexico/1388390

Redacción Publimetro. (2017, February 1). YouTube: ¿por qué Werevertumorro perdió cerca de medio millón de dólares? *Publimetro Perú.* https://publimetro.pe/redes-sociales/youtube-que-werevertumorro-perdio-cerca-medio-millon-dolares-55914-noticia/

Redacción Quién.com. (2014, April 3). Vlogueros dejan YouTube por polémica. *Quién.* https://www.quien.com/espectaculos/2014/04/03/vlogueros-dejan-youtube-por-polemica

Robertson, K. (2014). No One Would Murder For a Pattern: Crafting IP in Online Knitting Communities. En L. J. Murray, S. T. Piper, & K. Robertson (Eds.), *Putting Intellectual Property in its Place: Rights Discourses, Creative Labor, and the Everyday* (pp. 41–62). Oxford University Press.

Superholly. (2020, January 19). *MiTú se está robando nuestro dinero* [Youtube]. https://www.youtube.com/watch?v=5EtXkOqDkhg

Tabares, L. (2019). Professional Amateurs: Asian American Content Creators in YouTube's Digital Economy. *Journal of Asian American Studies, 22*(3), 387–417. https://doi.org/10.1353/jaas.2019.0029

Tu Cosmopolis. (2019, September 27). *La verdad de MI rancho a tu cocina.* https://www.youtube.com/watch?v=_Q4MhfcIbDo

Unocero Redacción. (2014, March 31). ¿Qué pasó con Werevertumorro? *Unocero.* https://www.unocero.com/noticias/que-paso-con-werevertumorro/

YouTubeTeam. (2007, December 10). Partner Program Expands. *Official YouTube Blog.* https://youtube.googleblog.com/2007/12/partner-program-expands.html

12 Mining as an art of survival in Venezuela

Eluding scarcity and improving living conditions with Bitcoins

David Ramírez Plascencia

Introduction

Since the publication of "Bitcoins: a peer to peer electronic cash system," the document which outlined the abstract and practical details of Bitcoins, in October 2008 by Satoshi Nakamoto (or the person or group behind him), the use of cryptocurrencies has quickly spread across the globe. People have become familiar with names like "Ethereum," "Ripple," and "Bitcoins." In 2018, the overall value of the global market of cryptocurrency surpassed $856 billion (GlobeNewswire, 2019), and by the end of 2019, there were about 2,788 different cryptocurrencies available to the public (CoinLore, 2019). In about a decade, cryptocurrencies have emerged as an option to acquire goods and services, and in the case of countries with authoritarian regimes or instable economies, it is one of the most trusted methods to buy and sell merchandises. However, this increasing popularity has fostered economic and legal concerns as well, such as the potentially negative impact of these currencies in the future of the global economy. Some experts argue that they are just financial bubbles that will eventually explode, generating a massive loss for investors and owners (Geuder et al., 2019; Patterson, 2019). In addition, there are ethical and legal concerns about the use of Bitcoins for illicit purposes. Indeed, Bitcoins, besides the latent devaluation problems, are very popular in dark net markets (Popper, 2020). This virtual cash are employed by criminal organizations such as narco gangs or terrorist groups to finance illegal businesses and operations: from the sale of drugs or weapons to the sponsor of extremist attacks. However, despite these negative views, cryptocurrencies provide huge advantages for the exchanging of common goods as well, particularly in countries with severe economic depression such as Venezuela: they are virtual and do not require a physical backup to operate, and they are more flexible than other traditional forms of payment, as they do not belong to a particular government or international organization. Therefore, cryptocurrencies are used to make money transfers that bypass financial controls, and the customer and the vendor could remain anonymous during the process, meaning they have less of a chance of being persecuted or having their funds seized by the government.

In Latin America, the popularization of cryptocurrencies is escalating. Every day the number of transactions in countries like Brazil and Argentina increases. In 2018, Argentina's capital city, Buenos Aires, was the place with the second most Bitcoin businesses in the world (Talty, 2018). Cryptocurrencies are very popular among Argentineans, since they represent an efficient form to protect their savings from the recurrent devaluation of the national currency the Peso (Szalay, 2019). Cryptocurrencies are seen as a method for surviving huge economic crises. Even the Venezuelan government has intended to create their own virtual coin "The Petro" whose value is sustained by the price of crude oil. Indeed, Venezuelans, in the country and abroad, are using virtual currencies to surpass official controls on foreign money, particularly the US dollar. In addition, they have established clandestine computer clusters to mine Bitcoin and obtain profits to help endure the difficult economic conditions in the country. Venezuelan migrants use Bitcoins to evade the financial system and send remittances to their relatives, to support their travel and accommodation during their journey into another country, and even to sell and buy houses in their homeland.

This chapter focuses on how Latin Americans, particularly Venezuelans have incorporated cryptocurrencies in their daily lives as "weapons" to fight back against economic control and overcome social deprivation, using bitcoins to improve their economic situation in their homeland and abroad. This study is based on the analysis of diverse real-world cases, legal codes, and economic reports covered by the press, analyzed in academic publications, or shared by Bitcoin users on social media. Section I focuses on how digital media and emerging technology as cryptocurrencies are used in the context of migration crises, not just by public authorities but migrants as well to support their families in their homeland; section II provides a general view about the legal and ethical concerns about cryptocurrencies, but how this technology could provide advantages for people living in difficult contexts as well. Section III shows some examples of how Latin Americans are using bitcoins in their daily life, and section IV centers on the analysis of in what way Venezuelans have incorporated bitcoins into their economy to surpass financial deprivation and scarcity, it provides some examples of how migrants use cryptocurrencies to support their families in the homeland as well.

Information technologies, migration crises, and the eruption of cryptocurrencies

At the beginning of the XXI century, migration crises are becoming more frequent, causing deep economic and political effects (Blondin & Boin, 2018). These emergencies stand as a great challenge for the host countries: the overload of humanitarian services and the social pressure from public opinion that demands the appropriate handling of the situation (Fontanari & Ambrosini, 2018). Yet, international and local media have exposed the deficiencies in infrastructure, logistics, and regulations on the part of local governments to respond suitably to the situation caused by the transit and accommodation of refugees, often entire families, in shelters and migratory stations. Migrants are usually housed in poor conditions, as has happened with Central Americans, Haitians, Cubans, Venezuelans, and even refugees from Africa in the United States and Mexico, or Sub-Saharans and Syrians in Europe and Turkey (Le Figaro Fr., 2019; Lissardy, 2019; Solera, 2019). In addition, the desperate situation of these people has become an opportunity for profit, not only for the smugglers, but sometimes even for public officers, as the case of *Cara di Mineo*, a refugee camp in Sicilia which closed in 2015, whose administration was investigated for corruption and misuse of public resources (Pianezzi & Grossi, 2018). Due to these challenges, authorities are trying to improve their response to migration emergencies through the use of technology.

New technologies as the case of drones or biometric devices have been used to establish surveillance systems to control the access to the host countries and to inhibit the entrance of terrorists and criminals (The Economist, 2019a), and in some cases, to manage and protect the refugee camps as well (Camacho et al., 2019). Drones, despite the high prices and the absence of regulations, are becoming popular not only for warfare purposes, but also to get strategic information and support the management of humanitarian crises. Drones are useful devices to monitor refugee camps that are susceptible to being attacked, to search for survivors from shipwrecks, or people lost in the desert. They could even be used to bring infrastructure and tools to hard-to-reach areas, or to provide internet connections to refugee camps (United Nations, 2017). Biometric technology is also employed to handle migration. Biometrics focuses on capturing the particular features of a person in order to determine their identity. One common usage of this technology is to assure the identity of visitors at the airports or the borders. In the case of refugee camps, biometrics serve to verify the use of resources allocated to each migrant (Camacho et al., 2019). Using iris recognition, stores will be able to confirm that the beneficiary is the person who is using the food or money card (Rahman et al., 2018; Johansson & Ljungek, 2019) The use of biometrics, however, as in the case of all other emerging technologies that gather personal information such as the case of drones, is not absent of legal concerns due to the potential risk of misusing private data. It requires public agencies to have close control over users' information, as there is sensitive data that may be used to defraud, steal, or extort migrants.

Recently, a new technology denominated Blockchain has gained global popularity. Blockchain is an information data structure in where the data is grouped into blocks. These blocks are aligned in such a way that the modification of a single line of code alters the entire chain, hence it is almost impossible to alter or falsify the data without noticing. For this reason, blockchain has been used as the basis to develop digital currencies like Bitcoin (Barber & Finley, 2019). The development of applications based on blockchain could play a significant role improving the lives of refugees by creating aid distribution systems that help migrants to pay for food and other basic goods with Bitcoins (Zambrano et al., 2018). Blockchain can provide additional advantages when dealing with the management of refugee camps, like protecting people's identifying data. Frequently, migrants lose passports, identity cards, and other important documents during their journey. With the implementation of a blockchain-based app that records their information, they could have an efficient way to protect their valuable data, not only from loss, but from being stolen by smugglers or confiscated by authorities (Huang, 2019). However, one of the most popular presence of the Blockchain technology among migrants is throughout the trade of Bitcoins to send remittances to their relatives at the homeland. But as it will be observed in the following section, this technology is not insusceptible to controversies, not just from the local governments in where cryptocurrencies avoid local financial controls and compromise local currencies but from international authorities that worry about the latent use of bitcoins to commit illicit activities.

Potentialities and concerns about cryptocurrencies

Since their irruption, virtual coins such as Bitcoin or Ethereum have fostered encouraging possibilities and deep concerns. They have been in the spotlight and under scrutiny by authorities because of their rising popularity in black markets hosted inside The Deep Web (Kethineni & Cao, 2019). In those places, Bitcoin serves as a common form of payment to trade illegal products like drugs, guns, or forbidden multimedia content (Rhumorbarbe et al., 2018). In addition, cryptocurrencies are employed to support international money laundering arrangements or to commit financial fraud in the form of the traditional Ponzi schemes (Campbell-Verduyn, 2017; Stokel-Walker, 2018). Also, of concern, it is their unstable value and lack of traditional financial security measures. On September 24, 2019, the price of Bitcoins fell by $1,000 in 30 minutes (The Economist, 2019c). Recently, QuadrigaCX, the largest Canadian cryptocurrency exchange enterprise, went bankrupt after the owner, Gerald William Cotton, who was the only person with access to the key to manage the funds, passed away in December 2018, leaving the investors in legal limbo. However, despite the controversies, many companies and even countries are considering the possibility of introducing novel currencies. This is the case of J.P. Morgan, Facebook, the Venezuelan government, and, recently in April 2020, the Chinese government announced that it is starting some tests to launch a new virtual currency (Cheng, 2020), though the confidence provided by these big-name adoptions does not seem to stem the concerns regarding the use of these virtual assets (Tang et al., 2019). Local and international authorities are less reluctant to investigate the exchange of cryptocurrencies and to establish novel regulations (Girasa, 2018; Herian, 2018; Kaiser et al., 2018). Yet, on the other side, blockchain technology that supports Bitcoin and Ethereum provides vast benefits when dealing with the safekeeping of personal data, the privacy of users, and the shield of transaction records (Francis, 2019).

Risen as a "libertarian response" due to the failure of central banks to tame the huge economic crisis of 2008 (Bariviera et al., 2017). The spread of Bitcoin could be seen as an effort to contest the influence of global financial organizations in local economies (Chalmers, 2018), and the materialization of an alternative decentralized form of social organization (Garrett et al., 2018). The object of blockchain technology is to settle on an alternative form of interchange, not just for economic purposes, but for a wide range of data, without intervention between the consumer and the supplier. The system allows all operations to remain anonymous since no personal data is needed to complete the transaction. In addition, cryptocurrencies do not rely on a unique central financial authority, but on a global network of about 52,500 computers running the same software (Barber & Finley, 2019). When a consumer wants to make a payment, an order is sent to the network, and the computers validate the action before the payment is registered in the blockchain's records (Low & Teo, 2017). The concrete foundation of this network is keeping an anonymous and accurate track of transactions and preventing the execution of any unauthorized modifications. Conversely, blockchain technology is more robust and versatile than just the development of cryptocurrencies, and its application extends far beyond the economic sphere. It could promote deep changes in the management of intellectual property like creating novel forms of distributing licensing rights, speeding up the payment of royalties (O'Dair, 2018), and preserving data integrity and authorship, some of the big concerns about the creation of digital content (Ito & O'Dair, 2018). Public authorities could implement online notary services based on blockchain to provide digital certificates, or to set electoral mechanisms that protect the integrity of votes (Datta, 2019), or to improve the logistics in the public agencies of big companies (Barber & Finley, 2019). In spite of the critics, cryptocurrencies are seen by some organizations, enterprises, and users as a promising technology that could provide financial services for those in complex situations, such as refugees or people living in deprived economies (Scott, 2016), enabling the sending of remittances or the establishment of alternative markets.

The art of survival with Bitcoins in Latin America

Despite the growing suspicions about virtual coins, Latin Americans have embraced cryptocurrencies as an alternative to traditional markets in order to bypass economic volatility and government monetary control. This is the case in Argentina, which for decades has suffered financial instability, huge inflation, and the recurrent devaluation of its national currency, the Peso. This has caused them to depend heavily on foreign currencies, particularly the US dollar, to stabilize its economy (Wilkis & Luzzi, 2019). However, in

2011, between the presidencies of Cristina Kirchner and current president Mauricio Macri, the US dollar became a less popular option, as the dollar's exchange rate was controlled by the government to protect the national currency (Moreno, 2016). In recent years, Bitcoin has gained popularity among Argentineans who would usually buy US dollars to protect their savings (Talty, 2018). Notwithstanding its volatility, Bitcoin has been a more stable choice than the national currency for acquiring goods and services (Popper, 2015), but particularly to protect their savings from the persistent devaluation of the Peso (Szalay, 2019). There are, however, still some legal shortfalls regarding cryptocurrencies in the country, since there is no law to protect people from losing their money, due to Bitcoin's instability and lack of a mechanism for surveilling users.

Other countries in the region have followed Argentina's example and moved toward implementing mechanisms to allow users to pay services with Bitcoin. In Rio de Janeiro, Brazil, it is possible to buy a public transportation ticket using Bitcoin (Rodrigues, 2019). Companies such as Amero-Isatek in Mexico are opening several exchange houses in cities such as Guadalajara, Mexico City, and Cancun. There have even been attempts to create a Mexican cryptocurrency, the Real Silver Coin, released by the company Sion Gold, whose value is backed by silver from the Mexican states of Durango and Sinaloa (Casas, 2019). Mexico is a pioneer in regulating cryptocurrencies or *activos virtuales* (digital assets) (El Heraldo de México, 2019), as they are named in the Mexican laws [Banco de México (Bank of Mexico), 2019]. According to the new regulation, the companies which want to operate in virtual coins must do so with authorization and must comply with requirements like providing information to customers about the possible risks of acquiring Bitcoins and protecting private client information. Additionally, they are under the scrutiny of the National Banking and Securities Commission (CNBV), an independent agency of the Secretariat of Finance and Public Credit (Mexico) (SHCP). Even the few Latin American countries that have not taken definitive steps towards the regulation of cryptocurrencies have emitted warnings about acquiring and using cryptocurrencies (Camargo Rico, 2017). Besides these remarkable examples, no other country in the world has adopted cryptocurrencies as a determinant element in their social and economic life as Venezuela has. The environment in this country has converted virtual coins, not only to an economic alternative, but a way of life.

The Venezuelan case

Recent Venezuelan history has been shaped by political polarization, economic hyperinflation, and high levels of criminality. In 2019, Venezuela's murder rate was 60 per 100,000 habitants, the greatest in Latin America (Pascuali, 2020). In addition to violence, citizens have to contend with political repression and the consequences of an economic deprivation (Hanson, 2018). The country is suffering the negative effects of massive hyperinflation which has converted the Bolivar, Venezuela's official currency, into a worthless paper, and has produced high levels of unemployment, triggering the scarcity of basic goods like groceries, medicine, and medical supplies (Nugent, 2018). The living conditions in Venezuela have paralleled countries in active war like Syria and Yemen: high mortality, scarcity of materials in hospitals, electricity shortages, and even the return of extinct diseases like rabies and measles. Millions of people, no matter their economic status, have left the country. At the time of this writing, so far into 2019, about four million Venezuelans have fled adverse living conditions in their home country (Long, 2019). Out of that number, nearly three million have established themselves in neighboring countries like Colombia (1,408,000), Peru (861,000), Ecuador (330,000), Chile (288,000), and Brazil (179,000) (The Economist, 2019b). Those migrants have been forced to settle in very poor conditions, often living in improvised shelters without basic sanitary and safety requirements. Deprived of the proper permissions to get a regular job, they commonly labor in low-wage jobs without any rights and even participate in dangerous activities like prostitution in Mexico and Spain (El Debate, 2018; Wallen, 2019). Despite this adverse context, Venezuelan migrants persevere. They have taken advantage of digital technologies (social media, mobile devices, and applications) and integrated them as central elements in their life. They use the Internet to build international solidarity networks to support their lives abroad and to help their relatives in the homeland, and they have even incorporated cryptocurrencies to send remittances to their families in Venezuela, as a safer form to elude financial controls from the Venezuelan authorities (Hernández, 2019).

People living in Venezuela, despite the lack of IT infrastructure and the frequent electricity shortages, are enthusiastic about digital technologies. According to the latest statistics from 2017 (Chevalier, 2019), about 60% of the population is connected to the Internet (about 19.1 million). One of the most popular activities online is the use of social media, particularly Facebook and WhatsApp (Anderson & Silver, 2019), but they have a strong presence on Twitter as well (Hausmann et al., 2018). This fact is not surprising, seeing as these virtual spaces are vital not just for maintaining personal relationships, but for political and economic purposes as well. Thanks to Facebook and Twitter, Venezuelans in the country and abroad are able to participate, at least at a distance, in the political agenda (Salojärvi, 2017). They follow politicians, local organizations, and media on Twitter and Facebook to stay aware of the main social and political events. Mobile applications like WhatsApp provide them with supportive spaces to build family or community groups, to stay in contact with their culture and traditions (Lozano, 2018), and to organize public demonstrations. With the support of virtual platforms and smartphones, they are able to search for jobs, sell merchandise, or to offer their professional services as medical professionals and lawyers as well (Alencar, 2019). Many of those offers are announced in closed Facebook groups, where Venezuelans in the country and abroad support the generation of markets, not only to supply the everyday needs of those in the host country, but to relieve their relatives and friends at home. It is in this last function where cryptocurrencies have played a key role.

Cryptocurrencies arrived in Venezuela as a desperate solution to a slew of economic, political, and social problems. It was not only the polarization of the society and the confrontation between President Nicolas Maduro and the political opposition, who pushed for clean elections and the ending of government repression. The huge economic crisis had caused a drawn-out decay of the general social conditions in the country. Bitcoins, despite its volatility, is a more trustworthy option than the national currency, and a weapon to bypass the government's strict control of US dollars (Haesly, 2017). Virtual coins have been the answer to the scarcity of goods and funds. One of the biggest advantages of cryptocurrencies is that they can be acquired outside the country, using US dollars or Mexican pesos, and they can be redeemed in Venezuela (Johnson, 2019). Because these operations do not rely on traditional bank accounts or depend on the intervention of any public authority, people can easily get Bitcoin to buy special gift cards online and then buy groceries or medicine (England & Fratrik, 2018). There are even some stores in the country where it is possible to pay directly in Bitcoin, like the drugstore *Farmarket,* which uses the application Xpay. Start-up companies like Send (https://www.sendprotocol.com) facilitate the process. This company has helped to establish an alternative money network based on blockchain where people can transfer, store, pay, or exchange bolivars and dollars using the *Send*'s SDT tokens. *Send* has a growing list of more than 85,000 users, of which 60% are Venezuelans, with half living in the country and the other half abroad (Marvin, 2018). In addition, ATM machines have been installed in Cucuta, a Colombian city in the Venezuela-Colombian border, that is visited everyday by thousands of Venezuelans, some of who are trying to leave the country, and some of who are just trying to buy basic needs in Colombia and return to Venezuela. Venezuelans use this ATM to exchange their Bitcoin (BTC), Bitcoin Cash (BCH), or Litecoin (LTC) for Colombian pesos (Salvo, 2019).

The Venezuelan government has been a fan of cryptocurrencies as well. In 2018 they created a virtual coin, the "petro" as an attempt to bring more stability to the national economy, and to bypass the economic sanctions imposed by the US government (Popper & Herrero, 2020). However, shortly after having been launched, the US banned it. Despite this fact, the petro is still operating in 2020, or is at least still offered in the official website (https://www.petro.gob.ve). Another economic activity linked to cryptocurrencies is mining. Generating Bitcoins is a process that requires huge amounts of electricity. Some estimations state that registering a Bitcoin transaction requires more than 5,000 times as much energy as using a credit card (Fairley, 2017). However, this is not a big problem in Venezuela since electricity, just like gasoline, is subsidized by the government. In 2018, Venezuela was the cheapest country in which to mine Bitcoin, about $531 US dollars. This number is even more impressive considering that in South Korea, the most expensive place to mine in, the cost is $26,170 (Morris, 2018). Individuals mining Bitcoins at home could make about $500 US dollars in a country where the minimum monthly wage is about $7 dollars (Cifuentes, 2019). Another "particular" economic activity in Venezuela is "farming" virtual gold in online videogames. People spend many hours playing in virtual roleplaying videogames to extract "gold coins." Venezuelans usually sell this money to other gamers in special online marketplaces. At the end, that gold is used to acquire special weapons or magic poisons to use in those games. Gamers can earn about $40 dollars a month which means more than five times the minimum wage ($7.5) in Venezuela. Sometimes the farmers use that virtual gold to acquire Bitcoins as a safer way to protect their money from economic depression (The Economist, 2019). While mining bitcoins in Venezuela could be "technically legal," this activity is very restrictive, and the authorities constantly harass bitcoin miners (Lanz, 2019). Working as a "Bitcoin miner" in Venezuela is a dangerous job, since the Venezuelan government intends to monopolize the bitcoins' market in the country. Federal authorities are paying more attention to the homes or enterprises that consume high levels of electricity as a method to uncover and capture Bitcoin miners and the police organize raids to seize computers and catch the miners, some of which are sent to jail (Sigalos, 2017).

Conclusions

As President Nicolas Maduro's government is running out of options to keep a basic level of economic stability in the country; cryptocurrencies are becoming a key alternate source of revenues not just for citizens but for the government as well, which are now attempting to popularize their own virtual coin "the petro," trying to avoid economic sanctions from the US government. At the same time, the venezuelan government are trying to exert more control on the diverse economic activities related to the virtual coins such as "farming," which, despite the frequent power outages and the authorities' harassment, is still a popular activity. However, in the long run, the more the country depends on farming Bitcoins, the more the infrastructure will be reliant on it that could compromise the viability of this incipient economic activity (Rosales, 2019). The government is accelerating the use of the petro among Venezuelans. Maduro has ordered the national banks to adopt the petro as a unit. All banks, private and public, should show finance information in bolivars and petros. There are public services that people can only pay using petros as passports, visas, and apostilles. The government is looking to entice Venezuelan migrants and convince them to use petros instead of other cryptocurrencies. In 2019, the Venezuelan government launched a new online service, Patria (Homeland), which allows the exchange of dollars for petros or bolivars. However, these attempts to popularize the petro are met with suspicion, not just because it has already been banned by the US, and not because it is very hard to find companies that are willing to operate in it, but because of the general distrust towards the Venezuelan government and its capacity (or lack thereof) to back the value of the cryptocurrency with crude oil (Gozzer, 2019). This will become even more complex during 2020, since the actual price of oil has collapsed because of the COVID-19 pandemic. Other external factors, such as international controls and surveillance of cryptocurrencies will eventually limit the use of these currencies in the country.

However, in the short term, Bitcoins will continue playing a key role. Its popularity in countries like Venezuela or Argentina show how new technologies could stand as crucial elements to improve people's living conditions. In Venezuela, Bitcoins have become a trustworthy alternative option to protect savings, send or receive remittances, pay for products and services, and even work, "farming" cryptocurrencies. Bitcoins have allowed migrants to support their families at the homeland by sending remittances to buy food or medicines avoiding the financial control of the Venezuelan government. In addition, these virtual coins have provided a safer method to protect their savings from the recurrent depreciation of the bolivar. Stores, companies, and even fast-food restaurants have found in Bitcoins a reliable way to stay in business in spite of hyperinflation. Despite their criticisms and their volatility, they have been incorporated into Venezuelans' daily life as weapons to combat scarcity and master the art of survival, in the country and abroad.

References

Alencar, A. (2019). Digital Place-Making Practices and Daily Struggles of Venezuelan (Forced) Migrants in Brazil. En *Handbook of Media and Migration*. Sage.

Anderson, M., & Silver, L. (2019, marzo 7). 7 key findings about mobile phone and social media use in emerging economies. *Pew Research Center.* https://www.pewresearch.org/fact-tank/2019/03/07/7-key-findings-about-mobile-phone-and-social-media-use-in-emerging-economies/

Banco de México (Bank of Mexico). (2019). *CIRCULAR 4/2019 «Activos Virtuales»* [DOF - Diario Oficial de la Federación]. Banco de México (Bank of Mexico). https://www.dof.gob.mx/nota_detalle.php?codigo=5552303&fecha=08/03/2019

Barber, G., & Finley, K. (2019, julio 9). Blockchain: The Complete Guide. *Wired.* https://www.wired.com/story/guide-blockchain/

Bariviera, A. F., Basgall, M. J., Hasperué, W., & Naiouf, M. (2017). Some stylized facts of the Bitcoin market. *Physica A: Statistical Mechanics and its Applications, 484*, 82–90. https://doi.org/10.1016/j.physa.2017.04.159

Blondin, D., & Boin, A. (2018). Managing Crises in Europe: A Public Management Perspective. En E. Ongaro & S. Van Thiel (Eds.), *The Palgrave Handbook of Public Administration and Management in Europe* (pp. 459–474). Palgrave Macmillan UK. https://doi.org/10.1057/978-1-137-55269-3_24

Camacho, S., Herrera, A., & Barrios, A. (2019). Refugees and Social Inclusion: The Role of Humanitarian Information Technologies. En S. Villa, G. Urrea, J. A. Castañeda, & E. R. Larsen (Eds.), *Decision-making in Humanitarian Operations: Strategy, Behavior and Dynamics* (pp. 99–123). Springer International Publishing. https://doi.org/10.1007/978-3-319-91509-8_5

Camargo Rico, L. Y. (2017). El efecto Bitcoin en la economía colombiana. Bogotá: Universidad Militar Nueva Granada.

Campbell-Verduyn, M. (Ed.). (2017). *Bitcoin and Beyond (Edición: 1)*. Routledge.

Casas, M. (2019, febrero 5). Crean en Durango criptomoneda respaldada con plata. *El Financiero*. https://www.google.com/search?client=safari&rls=en&q=Crean+en+Durango+criptomoneda+respaldada+con+plata&ie=UTF-8&oe=UTF-8

Chalmers, R. (2018). *The Politics Of Cryptography: How Has Cryptography Transformed Power Relations Between Citizens And The State Through Privacy & Finance?* [Master thesis, Leiden University]. https://openaccess.leidenuniv.nl/handle/1887/65157

Cheng, J. (2020, abril 20). China Rolls Out Pilot Test of Digital Currency. *The Wall Street Journal*. https://www.wsj.com/articles/china-rolls-out-pilot-test-of-digital-currency-11587385339

Chevalier, S. (2019). *Venezuela: Internet penetration 2017*. Statista. https://www.statista.com/statistics/209115/venezuela-internet-penetration/

Cifuentes, A. F. (2019). Bitcoin in Troubled Economies: The Potential of Cryptocurrencies in Argentina and Venezuela. *Latin American Law Review, 03*. https://doi.org/10.29263/lar03.2019.05

CoinLore. (2019). *List of All Cryptocurrencies*. CoinLore. https://www.coinlore.com/all_coins

Datta, A. (2019). Blockchain in the Government Technology Fabric. *arXiv:1905.08517 [cs]*. http://arxiv.org/abs/1905.08517

El Debate. (2018, julio 28). Las escorts que llegaron a México por un sueño y fueron asesinadas. *El Debate*. https://www.debate.com.mx/mexico/escorts-asesinadas-mexico-feminicidios-servicios-de-compania-20180728-0013.html

El Heraldo de México. (2019, septiembre 5). La guía básica para entender la Ley FINTECH. *El Heraldo de México*. https://heraldodemexico.com.mx/mer-k-2/la-guia-basica-para-entender-la-ley-fintech/

England, C., & Fratrik, C. (2018). Where to Bitcoin? *Journal of Private Enterprise, 33*(Spring 2018), 9–30.

Fairley, P. (2017). Blockchain world—Feeding the blockchain beast if bitcoin ever does go mainstream, the electricity needed to sustain it will be enormous. *IEEE Spectrum, 54*(10), 36–59. https://doi.org/10.1109/MSPEC.2017.8048837

Fontanari, E., & Ambrosini, M. (2018). Into the Interstices: Everyday Practices of Refugees and Their Supporters in Europe's Migration 'Crisis'. *Sociology, 52*(3), 587–603. https://doi.org/10.1177/0038038518759458

Francis, J. C. (2019). *Bitcoins, Cryptocurrencies and BlockChains* (SSRN Scholarly Paper ID 3371051). Social Science Research Network. https://papers.ssrn.com/abstract=3371051

Garrett, M., Catlow, R., Skinner, S., & Jones, N. (Eds.). (2018). *Artists Re:thinking the Blockchain* (Edición: 1). Liverpool University Press.

Geuder, J., Kinateder, H., & Wagner, N. F. (2019). Cryptocurrencies as financial bubbles: The case of Bitcoin. *Finance Research Letters, 31*. https://doi.org/10.1016/j.frl.2018.11.011

Girasa, R. (2018). *Regulation of Cryptocurrencies and Blockchain Technologies: National and International Perspectives* (Edición: 1st ed. 2018). Palgrave Macmillan.

GlobeNewswire. (2019, septiembre 9). The World Market for Cryptocurrency. *Business Insider*. https://markets.businessinsider.com/news/stocks/the-world-market-for-cryptocurrency-2017-2018-review-2019-2024-forecast-with-analysis-on-bitmain-technologies-bitgo-nvidia-corporation-ripple-networks-and-coinbase-1028508056

Gozzer, S. (2019, julio 24). *Qué fue del petro, la criptomoneda con la que Venezuela quería evadir las sanciones económicas.* BBC News Mundo. https://www.bbc.com/mundo/noticias-america-latina-49045096

Haesly, K. (2017). How to Solve a Problem Like Venezuela: An Argument for Virtual Currency. *Law and Business Review of the Americas, 22*(3), 261.

Hanson, R. (2018). Deciphering Venezuela's Emigration Wave. *NACLA Report on the Americas, 50*(4), 356–359. https://doi.org/10.1080/10714839.2018.1550976

Hausmann, R., Hinz, J., & Yildirim, M. A. (2018). *Measuring Venezuelan emigration with Twitter* (Working Paper N.o 2106). Kiel Working Paper. https://www.econstor.eu/handle/10419/179127

Herian, R. (2018). *Regulating Blockchain: Critical Perspectives in Law and Technology* (Edición: 1). Routledge.

Hernández, C. (2019, febrero 23). Bitcoin Has Saved My Family. The New York Times. https://www.nytimes.com/2019/02/23/opinion/sunday/venezuela-bitcoin-inflation-cryptocurrencies.html

Huang, R. (2019, enero 27). How Blockchain Can Help With The Refugee Crisis. *Forbes.* https://www.forbes.com/sites/rogerhuang/2019/01/27/how-blockchain-can-help-with-the-refugee-crisis/#43eba7f65621

Ito, K., & O'Dair, M. (2018). A Critical Examination of the Application of Blockchain Technology to Intellectual Property Management. En H. Treiblmaier & R. Beck (Eds.), *Business Transformation through Blockchain: Volume II* (Edición: 1, pp. 317–335). Palgrave Macmillan.

Johansson, K., & Ljungek, F. (2019). *Global Solution, Local Inclusion?: A study of digital IDs for refugees in Uganda* [Uppsatser - Kulturgeografiska institutionen, Stockholms universitet]. http://urn.kb.se/resolve?urn=urn:nbn:se:uu:diva-385681

Johnson, J. (2019). Bitcoin and Venezuela's Unofficial Exchange Rate. *Ledger, 4*(0). https://doi.org/10.5195/ledger.2019.170

Kaiser, B., Jurado, M., & Ledger, A. (2018). The Looming Threat of China: An Analysis of Chinese Influence on Bitcoin. *arXiv:1810.02466 [cs].* http://arxiv.org/abs/1810.02466

Kethineni, S., & Cao, Y. (2019). The Rise in Popularity of Cryptocurrency and Associated Criminal Activity. *International Criminal Justice Review,* 1057567719827051. https://doi.org/10.1177/1057567719827051

Lanz, J. A. (2019, julio 27). Confessions of a Venezuelan Bitcoin miner. *Decrypt.* https://decrypt.co/8092/confessions-of-a-venezuelan-bitcoin-miner

Le Figaro Fr. (2019, febrero 6). Migrants: 34 ONG appellent l'ONU à venir à Calais. *Le Figaro Fr.* http://www.lefigaro.fr/flash-actu/2019/02/06/97001-20190206FILW-WW00275-migrants-34-ong-appellent-l-onu-a-venir-a-calais.php

Lissardy, G. (2019, junio 27). Por qué están llamando "campos de concentración" a los lugares de detención de inmigrantes en Estados Unidos. *BBC News Mundo.* https://www.bbc.com/mundo/noticias-internacional-48781955

Long, G. (2019, agosto 24). Venezuelan refugee exodus intensifies. *Financial Times.* https://www.ft.com/content/fe0291a2-c5ab-11e9-a8e9-296ca66511c9

Low, K. F., & Teo, E. G. (2017). Bitcoins and other cryptocurrencies as property? *Law, Innovation and Technology, 9*(2), 235–268. https://doi.org/10.1080/17579961.2017.1377915

Lozano, D. (2018, diciembre 24). Celebran venezolanos Navidad en WhatsApp. *Reforma.* https://www.reforma.com/aplicacioneslibre/preacceso/articulo/default.aspx?id=1570650&v=3&urlredirect=https://www.reforma.com/aplicaciones/articulo/default.aspx?id=1570650&v=3

Marvin, R. (2018, agosto 9). In Venezuela, Cryptocurrency Is an Oppressor and a Lifeline | PCMag.com. *PC Magazine*. https://www.pcmag.com/feature/362486/in-venezuela-cryptocurrency-is-an-oppressor-and-a-lifeline

Moreno, E. (2016). Bitcoin in Argentina: Inflation, Currency Restrictions, and the Rise of Cryptocurrency. *International Immersion Program Papers*. https://chicagounbound.uchicago.edu/international_immersion_program_papers/18

Morris, C. (2018, marzo 7). Here's What It Costs to Mine a Bitcoin Around The World | Fortune. *Fortune*. https://fortune.com/2018/03/07/bitcoin-mining-costs-global-south-korea-venezuela/

Nugent, C. (2018, octubre 23). How Hunger Fuels Crime and Violence in Venezuela. *Time*. http://time.com/longform/hunger-crime-violence-venezuela/

O'Dair, M. (2018). *Distributed Creativity: How Blockchain Technology will Transform the Creative Economy* (Edición: 1). Palgrave Macmillan.

Pascuali, M. (2020). *Homicide rates in Latin America & the Caribbean by country 2019*. Statista. https://www.statista.com/statistics/947781/homicide-rates-latin-america-caribbean-country/

Patterson, M. (2019, abril 4). Has Bitcoin Bottomed? Here's How It Compares With Past Bubbles. *Bloomberg.Com*. https://www.bloomberg.com/news/articles/2019-04-04/has-bitcoin-bottomed-here-s-how-it-compares-with-past-bubbles

Pianezzi, D., & Grossi, G. (2018). Corruption in migration management: A network perspective. *International Review of Administrative Sciences*, 0020852317753528. https://doi.org/10.1177/0020852317753528

Popper, N. (2015, abril 29). Can Bitcoin Conquer Argentina? *The New York Times*. https://www.nytimes.com/2015/05/03/magazine/how-bitcoin-is-disrupting-argentinas-economy.html

Popper, N. (2020, enero 28). Bitcoin Has Lost Steam. But Criminals Still Love It. *The New York Times*. https://www.nytimes.com/2020/01/28/technology/bitcoin-black-market.html

Popper, N., & Herrero, A. V. (2020, marzo 20). The Coder and the Dictator. The New York Times. https://www.nytimes.com/2020/03/20/technology/venezuela-petro-cryptocurrency.html

Rahman, Z., Verhaert, P., & Nyst, C. (2018). *Biometrics in the Humanitarian Sector*. Oxfam/The Engine Room.

Rhumorbarbe, D., Werner, D., Gilliéron, Q., Staehli, L., Broséus, J., & Rossy, Q. (2018). Characterising the online weapons trafficking on cryptomarkets. *Forensic Science International*, *283*, 16–20. https://doi.org/10.1016/j.forsciint.2017.12.008

Rodrigues, L. (2019, agosto 1). Agora é possível pagar o metrô do Rio de Janeiro com cartão Visa carregado com Bitcoin. *CriptoFácil*. https://www.criptofacil.com/agora-e-possivel-pagar-o-metro-do-rio-de-janeiro-com-cartao-visa-carregado-com-bitcoin/

Rosales, A. (2019). Radical rentierism: Gold mining, cryptocurrency and commodity collateralization in Venezuela. *Review of International Political Economy*, *0*(0), 1–22. https://doi.org/10.1080/09692290.2019.1625422

Salojärvi, V. (2017). The Media Use of Diaspora in a Conflict Situation: A Case Study of Venezuelans in Finland. En O. Ogunyemi (Ed.), *Media, Diaspora and Conflict* (pp. 173–188). Palgrave Macmillan. https://doi.org/10.1007/978-3-319-56642-9_11

Salvo, M. D. (2019, marzo 19). *Why are Venezuelans seeking refuge in crypto-currencies? BBC News*. https://www.bbc.com/news/business-47553048

Scott, B. (2016). *How can cryptocurrency and blockchain technology play a role in building social and solidarity finance?* (Working Paper N.o 2016-1). UNRISD Working Paper. https://www.econstor.eu/handle/10419/148750

Sigalos, M. (2017, septiembre 2). *This is one of the world's most dangerous places to mine bitcoin. CNBC.* https://www.cnbc.com/2017/08/30/venezuela-is-one-of-the-worlds-most-dangerous-places-to-mine-bitcoin.html

Solera, C. (2019, junio 28). Administrativos cuidan a migrantes; señalan que los pusieron en riesgo. *Excélsior.* https://www.excelsior.com.mx/nacional/administrativos-cuidan-a-migrantes-senalan-que-los-pusieron-en-riesgo/1321337

Stokel-Walker, C. (2018). The murky world of the bitcoin scam. *New Scientist, 237*, 12. https://doi.org/10.1016/S0262-4079(18)30060-5

Szalay, E. (2019, mayo 27). Bitcoin hits all-time high in Argentine pesos. *Financial Times.* https://www.ft.com/content/44bfb5b8-7d4d-11e9-81d2-f785092ab560

Talty, A. (2018, julio 31). The Top 10 Bitcoin Cities In The World. *Forbes.* https://www.forbes.com/sites/alexandratalty/2018/07/31/the-top-10-bitcoin-cities-in-the-world/#79cfa84c4565

Tang, Y., Xiong, J., Becerril-Arreola, R., & Iyer, L. (2019). Blockchain Ethics Research: A Conceptual Model. *Proceedings of the 2019 on Computers and People Research Conference*, 43–49. https://doi.org/10.1145/3322385.3322397

The Economist. (2019a, febrero 14). Technology could make a hard border disappear, but at a cost. *The Economist.* https://www.economist.com/science-and-technology/2019/02/14/technology-could-make-a-hard-border-disappear-but-at-a-cost

The Economist. (2019b, septiembre 12). Millions of refugees from Venezuela are straining neighbours' hospitality. *The Economist.* https://www.economist.com/the-americas/2019/09/12/millions-of-refugees-from-venezuela-are-straining-neighbours-hospitality

The Economist. (2019c, octubre). Betting on bitcoin prices may soon be deemed illegal gambling. *The Economist.* https://www.economist.com/finance-and-economics/2019/10/03/betting-on-bitcoin-prices-may-soon-be-deemed-illegal-gambling

The Economist. (2019, noviembre 21). Venezuela's paper currency is worthless, so its people seek virtual gold. *The Economist.* https://www.economist.com/the-americas/2019/11/21/venezuelas-paper-currency-is-worthless-so-its-people-seek-virtual-gold

United Nations. (2017). FEATURE: Does drone technology hold promise for the UN? *Refugees and Migrants.* https://refugeesmigrants.un.org/feature-does-drone-technology-hold-promise-un

Wallen, J. (2019, febrero 1). How Venezuela's crisis is fuelling prostitution and sex trafficking on Spain's Costa del Sol. *The Telegraph.* https://www.telegraph.co.uk/global-health/women-and-girls/venezuelas-crisis-fuelling-prostitution-sex-trafficking-costa/

Wilkis, A., & Luzzi, M. (2019). *El dólar.* Crítica Argentina.

Zambrano, R., Young, A., & Verhulst, S. (2018). *Connecting Refugees to Aid through Blockchain- Enabled ID Management: World Food Programme's Building Blocks.* Govlab.

13 Conclusions

Avery Plaw, Barbara Carvalho Gurgel,
and David Ramírez Plascencia

A defiant context

The rise of China as an economic and military superpower has created tensions and misgivings with other well-established hegemonies such as the United States, the European Union, and Japan, who now reluctantly observe how the Asian giant has expanded their presence in Africa, the Middle East, and Latin America (Carneiro Corrêa Vieira, 2019). In recent years, China has become a great investor and lender for Latin American countries like Brazil, Ecuador, Argentina, and Venezuela (Vila Moreno, 2020). These actions have granted China political influence and potential control over strategic raw materials like oil, copper, and soybeans. There are other sectors, however, such as technology and telecommunications in which China has exponentially advanced. It is precisely in those areas where this distrust is becoming more evident (Segal, 2019). Former economic alliances among US and Chinese corporations, where, for example, the product was designed in the Silicon Valley and then made in China, are now under public scrutiny. The US government has blamed China for using companies such as Huawei, the world's largest telecom equipment-maker, to spy and steal business secrets. Consequently, the US government started a trade war that involved the imposition of tariffs, export restrictions and the settlement of a wary media campaign towards China. Finally, after many hard conversations, both countries signed a truce on January 15, 2020. They officially agreed to the rollback of taxes, and to increase trade.

This recent case illustrates how, despite the US' high-tech supremacy, the Asian country is no longer a secondary player that just manufactures products designed in the US. Chinese companies such as Alibaba or Tencent have a market value similar to Facebook (The Economist, 2018) and the country is a leader in the development of artificial intelligence and 5G technology. Beyond the relevance of this trade divergence for the IT sector, the US-Chinese conflict educates us about the prominence of digital technologies in the 21st century, not just for economic or social purposes but, as smartphones, drones, and artificial intelligence are incorporated into people's daily lives, their development and control is becoming crucial for homeland security as well. Superpowers are not only concerned about producing new missiles or jets, but they are struggling to consolidate them as "cyber behemoths" that can surveil and control internet flows, and even to sabotage and alter other countries' key services that depend on information technologies such as airports or subways (O'Flaherty, 2018). Under these circumstances, the equilibrium between surveillance and privacy has become one the most important matters in modern democracies. On one side, there is the State's duty of protecting and securing citizen lives and possessions. On the other, there are huge apprehensions about how authorities could use technology to control people's opinions and censure critics (Yuan, 2020).

Keeping this balance is particularly complex in circumstances of great social pressure, political polarization, and uncertainty about economic recovery after the COVID-19 pandemic (World Health Organization, 2020). Governments across the globe have to contend with a deadly disease that rapidly spreads among the population (Berkeley, 2020). In a matter of three months, since the first reported contagion in December 2019 in the city of Wuhan in the Chinese province of Hubei, the disease became global. Suddenly, hospitals in Spain and Italy became overwhelmed (Sevillano & Rincón, 2020). As the effects of the COVID-19 pandemic spread, authorities were forced to impose draconian procedures such as the ceasing of non-essential activities and strict isolation of people at home. Drones were used to secure urban and rural zones, and to guarantee observation of social ordinances. The Chinese government took a further step and employed geolocation technology in smartphones to control people's movements. But it was technology that allowed people to stay in contact with their social life in spite of the confinement. Thanks to applications like ZOOM and WhatsApp, students were able to continue their education at home, employers were able to work at home, and relatives could talk with their loved ones in hospitals. It was through social platforms that people collaborated in public campaigns to gather resources and deliver medical material. This extreme event let people experience the advantages and perils of the omnipresence of digital media in modern hyper-connected societies.

This first volume of the editorial project The Politics of Latin America provided remarkable examples about the impact of popular and emerging technologies in political, economic, and security issues. Chapters delivered insightful arguments on how the region, in spite of their institutional, legal, and high-tech shortfalls, is entering into a hyper-technologization trend. As we have seen in this volume, the settlement of open government systems, the use of big data, and the incorporation of drones and artificial intelligence into labor, health, and security affairs raise significant legal and ethical dilemmas in a region where democratic institutions and human rights are very fragile (Malaquias & Albertin, 2018). Also it was possible to observe across these chapters how digital media is no longer a possibility, but a reality in Latin America, not only regarding the popularity of social media and entertainment platforms such as Netflix and YouTube, but also with the quick popularization of drones, and the growing attention towards artificial intelligence and the growing presence of robots in the manufacturing sector.

However, this technological adoption is not immune to conflicts in a region where corruption, the abuse of power and censorship is persistent. Citizens are concerned about the possible backlash and potential negative use of technology by authorities under the argument of improving surveillance and prosecuting crime. Cases like Wikileaks and Edward Snowden have shown that governments are not reluctant to use digital technologies to spy on other countries and citizens (MacAskill, 2017). Those cases have brought the importance of setting limits to the surveillance of authorities on their citizens and the protection of civil rights on the Internet (Schuster et al., 2017) under the media spotlight. Moreover, in this volume we have observed in what way the eruption of digital media has offered advantages to criminal organizations to expand and diversify their operations. Narco gangs in Mexico and Colombia have built international distribution networks, increasing their influence and power far beyond national borders. This conclusion will discuss three main issues: (i) how public institutions and corporations face ethical and legal critical complexities related to data protection and privacy, (ii) in what way the incorporation of robots, drones, and artificial intelligence to combat organized crime will eventually impact the protection of civil rights, and (iii) what could be the potential outcomes of the digital transformation of the economy in Latin American?

Public affairs and the ethical use of private data

As the popularization of digital media grows across the developing world, millions of new customers are opening email accounts and sharing information on their social media profiles. These platforms offer them new political and social possibilities: to stay in contact with their relatives and friends abroad, get information, engage in public affairs, and search and apply for jobs. In exchange, these platforms gather information about their preferences, tastes, and opinions, then they sell that data to marketing agencies and corporations. An important share of Google and Facebook's revenues rely on the information generated by their users (Lehuedé, 2019). At a simple glimpse, this seems a fair exchange. Both parts get benefits from the use of social media. However, cases such Cambridge Analytica show that companies are more interested in making profits than protecting people's data. Since this scandal, there has been a rising concern about improving the protection of private data not only from a potential misuse from governments and corporations, but of scams and frauds online. Improving digital privacy is critical regarding children and young ones which are exposed to diverse harms when browsing online: grooming, cyberbullying, or the incidental exposure to sexual and violent content. Consequently, in recent years, big media corporations like Facebook have been required to implement new tools and directives to keep their users from a potential misuse of their private information (Todt, 2019). In early 2020, this company launched a new section inside the user settings which allows people to see which third-parties have distributed consumer activities with the social network, in addition people can use this tool to delete their "history" (a recount of past activities inside the platform) from their social media profile (Fowler, 2020).

Media corporations, however, are not the only sector that are greatly interested in private data. Information is crucial for the government to set proper social policies, recollect taxes, and deliver enhanced public services. But in order to make an efficient use of data, governments, particularly in Latin America, require the establishment of novel digital systems which are able to comply with the highly unequal context in where they will operate. In addition, it is important the settlement of inclusive guidelines as the use of assistive technologies for people with disabilities and the employment of indigenous languages in the menu of applications to encourage access of ethnic minorities to public services. Furthermore, those actions demand the design of policies that promote an equilibrium among individual and community interests about data, the right of privacy, and the government's requests to collect and manage citizens' information. These actions are particularly mandatory in regions like Latin America where a big segment of the population lack the proper digital literacy and technological equipment to properly access digital services. In addition, it is mandatory to set robust legal frameworks related with the gathering, use, and distribution of public and private data (Greenleaf, 2019). In this context, countries such as Mexico and Colombia have promulged recent norms on the subjects. Those codes establish frames and procedures about how corporations and governments are responsible to protect the private data of their users/citizens (Silva et al., 2019; Mendoza Enríquez, 2018).

Other technologies employed by pubic authorities to collect people's information, such as QR cards and biometrics, face great challenges as well, particularly about their incorporation to control immigration or to deliver public services. The Venezuelan government has implemented the use of QR cards to centralize citizens' socioeconomic information. Called the Homeland card (Carnet de la Patria), this ID is employed by the authorities to distribute social program aids: food, health services, and access to subsidized gasoline (Berwick, 2018). At the same time, this technology allows authorities to segregate and discriminate against people depending on if they support the current government or not. Biometric authentication implies potential risks as well: sensor inaccuracy, hacking, and misuses of biobanks. This concern not only applies in the context of political polarization as it happens in Venezuela, but in well-established democracies that have to deal with the administration of biometric technologies to control migration at national borders. In order to protect people's private information from possible abuses, it is important that the design of these systems is based on well-defined legal frameworks and the participation of corporations and civil organization along with the authorities (Díaz, 2014).

The protection of private data and the sponsorship of civic participation in the decision-making process is fundamental to develop inventive models of administration that could bring novel solutions to traditional problems such as criminality and social exclusion. In this situation, the open government model could be an essential tool to promote transparency and the arrangement of mechanisms that favor public scrutiny of the actions of authorities (Alzamil & Vasarhelyi, 2019). Its implementation, however, requires horizontal dialogue among citizens and the government, public accountability and transparency. In Latin America, the future prospects of this innovative form of government faces important social, political, and technological barriers: decreasing corruption, the improvement of infrastructure and connectivity, and positive governmental actions to recover people's trust in public institutions. For that reason, the conformation of civil organizations that promote the arrangement of legal and technological frameworks to map, gather data, and make independent reports on regional problems as crime and domestic violence is important (Cullen et al., 2019).

Homeland security and human rights, a problematic balance?

In 2020, there are about 4.54 billion internet users. Everyday activities such as reading newspapers, buying food or socializing occurs mostly on digital platforms and mobile applications. But digital media provides not only advantages but perils as well. Criminals have gained benefits from the Internet. They use technology to steal and scam people on a global scale. Illegal activities are varied and recurrent in cyberspace: cracking, fraud, the commerce of contraband, virtual kidnapping, and even the use of social media for human trafficking. Prosecuting cybercrime is difficult since many of these activities take place across several national borders and jurisdictions (Shillito, 2019). In Latin America, as the number of users, portals, and services flourish, cybercrime is becoming more frequent (Kshetri, 2013). Factors such as the inexperience of users and corporations, the lack of protocols to combat these actions and the lack of efficient legal frameworks exacerbate the problem (Fernández, 2019). Inexperience using online services causes the exposure of sensitive data such as passwords or credit card numbers, in many cases, incipient small businesses and public departments are not implementing enough security protocols and technical frameworks to protect the user's information (Michalczewsky & Ramos, 2017). These deficiencies allow the occurent of novel forms of law-breaking in where criminals take advantage of digital media not only to steal or scam people's money, but to commit sexual exploitation, abuse, and traffic humans (Sarkar, 2015). The diversification of digital media in Latin America has brought a new abundance of sexual abuse: sextortion, grooming, and child pornography distribution. These illegal practices have proliferated partly because of the omnipresence of technology in children and young people's lives who spend many hours per day using social media. In this sense, it becomes imperative to set efficient prevention plans and legal frameworks to secure the children's experiences (OAS, 2018).

While Latin American countries struggle to set integral solutions to protect their citizens from online crime, emerging technologies such as drones are booming along with a growing interest in robots and the development of applications based on artificial intelligence. Regarding homeland security, drones have been quickly adopted to improve surveillance in the urban context where infrastructure prevents police force access. However, the use of drones to combat criminality is controversial since it could violate the right to privacy of the people, who could be eventually recorded and incriminated without any legal reserve. Drones have been used by civil organizations, the government and the criminal gangs in the drug war in Mexico. Every actor has gained significant advantages from these devices for surveilling remote areas, to search for victims' burials in remote rural areas, and in the case of criminal gangs to deliver drugs across the border, or even attack police officers. In the US-Mexican border, narcodrones represent a new face in a very long conflict among the authorities and the narcos. In this context, drones evoke ideas of technical sophistication. But at the same time, the presence of narcodrones has been capitalized by authorities on both sides of the border to institutionalize a state of exception, allowing them to undertake illegal and unethical actions. In the forthcoming years, the success or failure of the incorporation of drones, and other novel technologies as lethal autonomous weapon systems (LAWS), into security strategies not just in Mexico, but across the region, will depend on the harmonization of legal a framework and the improvement of horizontal dialogue among the civil organizations and authorities.

Labor Markets and emerging technologies' promises and threats

The huge impact of digital media in Latin America can be clearly seen in the influx and rise of electronic commerce: companies such as Amazon and Mercado Libre are very popular options to sell and acquire goods. In many cities, people can use Uber for going to school or work and even to buy food in restaurants. But entertainment has changed as well, since the entrance of Netflix in 2011, other companies such HBO and Disney have developed applications for mobile phones and Smart TVs to deliver their content to a growing market that has shifted from traditional forms of entertainment to the consumption of digital content. Video Streaming (SVoD) is growing exponentially in the region. Watching movies and series online is a popular pastime among Latin Americans. One vivid example of this trend could be found in the phenomenal growth of the user-generated content market. This rising entertainment is supported by millions of users who are very enthusiastic about digital content, mostly videos, created and shared on social media by users known as YouTubers. Latin American YouTuber profiles have millions of followers, and their annual revenues are counted in the millions of US dollars.

In this scenery where people are moving towards digital media, other economic technologies such as cryptocurrencies, robots, and artificial intelligence applications are entering into the Latin American economy as well. People in countries like Argentina and Venezuela, for example, have started using bitcoins to save their money due to the precarity of their national currencies. Despite the rising popularity of cryptocurrencies in Latin America and the optimism shared among users and companies that sell and use them, there are growing concerns among international financial authorities about the potential use of these currencies to back criminals and terrorist organizations. Hence, there will be more controls and surveillance towards the production and distribution of bitcoins, but, in the meantime, those instruments will continue playing a key role for people living in countries with deprived economies.

As we have observed in this volume, fresh technologies, such as drones, are gaining great attention among governments and the public as well. Other devices and applications, like robots and AI applications look encouraging. Despite the fact that robots have had a very long history, testimonies about the construction of automatons could possibly be dated to the ancient Greeks and certainly to Leonardo da Vinci (Mayor, 2018), their importance has vastly increased in recent decades. In addition, detailed depictions, expectations, and fears regarding robots could be found in sci-fi works like Metropolis and 2001: A Space Odyssey or I, Robot. These movies well depict the potential and perils traditionally associated with robots, especially when humans endow them with the ability to "think for themselves," losing, at the same time, the power to control the robots' actions. This could explain, from a cultural point of view, the latent human fear of an eventual robot revolution, as depicted in diverse fictional works. But far beyond science fiction, the risk of labor substitution by robots and the automatization of production in several industries is latent. A major presence of robots to undertake human labor could compromise most low-skill jobs in manufacturing companies. In this context, it is important to set policies that incentivize a regional economy that depends less on manufacturing for the development of qualified jobs based on the development of innovation and the production of digital assets.

It is because of these backgrounds that the presence for robots in industries is promising but diverging at the same time in a region where many people work in manufacturing sectors. There are many concerns about the potential loss of jobs due to the entrance of robots. Many people believe they could eventually replace humans at factories making cars, one of the biggest activities in Mexico and Brazil. Governments in Latin America tend to use emerging technologies, like drones, robots, or AI, to set narratives of progress and development. But it is important to take a more assertive path to incentivize research into the development of AI applications and robots that could bring novel solutions to traditional problems in great urbanizations, such as pollution, traffic, and crime. In this context, the development of well-defined and integral legal frameworks is mandatory. Convincing and well-defined normativity will provide security to investors and researchers about the protection of their intellectual property rights and will set clear limits to the legal responsibility of the developers of future AI systems. Doing so will bring harmony for both the users and the companies that create these technologies.

Final remarks

As governments are debating how to properly allow the reopening of commerce and industry following the COVID-19 pandemic, the reducing of seclusion ordinances, and the returning to school for thousands of children, there are still concerns about the occurrence of new outbreaks and the final economic consequences of the pandemic: massive loss of jobs, severe public spending cuts, and the growth of poverty and social exclusion. The Latin American context, shaped by low wages and precarious labor conditions, ventures a pessimistic view in the short run. Despite the efforts to diminish the effect of the social lockdown, some sources estimate that the regional economy will shrink 7.2% in 2020 (The Economist, 2020). Hence, many Latin American governments are trying to recover economic activities without even reaching the peak of the pandemic. Returning to the "new normal" in a region with high levels of social inequality will be challenging.

To surpass this, individuals, corporations, and public agencies have incorporated technology to "virtualize" their processes and activities. Digital media has served to mediate personal communication when the new ordinances prescript social distancing and the avoidance of gatherings and physical contact. Despite that, some authorities have reduced or eliminated these dispositions. There are doubts about how social distancing could apply when dealing with overloaded public transport or marketplaces as is happening in some metropolises in Mexico and Brazil, two countries with the most COVID-19 casualties in the region. In these cases, technology will still play a key role by providing alternative forms to buying groceries or working and studying at home. During these months, new prototypes and applications, based on for example AI and Bluetooth connectivity, have been developed to prevent and detect infections without using invasive methods. As much as these technologies could look promising, it is important that all researches about COVID-19 and other mortal diseases are conducted under strict protocols not just to guarantee the effectiveness of the tests and treatments but to properly protect the digital and bio data of the individuals who participate in these trials.

Probably one of the most defiant issues that societies must face when dealing with very adverse contexts as in the case of war, pandemic, or natural disasters, is how to avoid the temptation of sacrificing civil rights for security and protection. History provides plenty of examples where those calamities are used by unscrupulous rulers as an excuse to limit citizen rights and expand government reach. The 2020 COVID-19 crisis is not an exception. Cases in Hungry, El Salvador, and Kenya confirm this fact. In this actual hyper-connected context, security and protection should be an imperative, not just in the streets but particularly in cyberspace which has been growing as a key component for diverse social, political, and economic activities. It is mandatory in this case to clarify legal restrictions in between what can permissibly be used for economic, security, and public interests (which have notoriously been abused by states) and what must remain private and anonymous as part of the undeniable domain of civil rights (which remains dangerously fragile).

References

Alzamil, Z. S., & Vasarhelyi, M. A. (2019). A new model for effective and efficient open government data. *International Journal of Disclosure and Governance, 16*(4), 174–187. https://doi.org/10.1057/s41310-019-00066-w

Berkeley, L. Jr. (2020, March 26). The coronavirus may be deadlier than the 1918 flu: Here's how it stacks up to other pandemics. *CNBC.* https://www.cnbc.com/2020/03/26/coronavirus-may-be-deadlier-than-1918-flu-heres-how-it-stacks-up-to-other-pandemics.html

Berwick, A. (2018, November 14). A new Venezuelan ID, created with China's ZTE, tracks citizen behavior. *Reuters.* https://www.reuters.com/investigates/special-report/venezuela-zte/

Carneiro Corrêa Vieira, V. (2019). From Third World Theory to Belt and Road Initiative: International Aid as a Chinese Foreign Policy Tool. *Contexto Internacional, 41*(3), 529–551. https://doi.org/10.1590/s0102-8529.2019410300003

Cullen, P., Vaughan, G., Li, Z., Price, J., Yu, D., & Sullivan, E. (2019). Counting Dead Women in Australia: An In-Depth Case Review of Femicide. *Journal of Family Violence, 34*(1), 1–8. https://doi.org/10.1007/s10896-018-9963-6

Díaz, V. (2014). Legal challenges of biometric immigration control systems. *Mexican Law Review, 7*(1), 3–30. https://doi.org/10.1016/S1870-0578(16)30006-3

Fernández, A. (2019, June 14). Record number of online frauds in Mexico. *El Universal.* https://www.eluniversal.com.mx/english/record-number-online-frauds-mexico

Fowler, G. A. (2020, January 28). Perspective | Facebook will now show you exactly how it stalks you—Even when you're not using Facebook. *Washington Post.* https://www.washingtonpost.com/technology/2020/01/28/off-facebook-activity-page/

Greenleaf, G. (2019). *Personal Data Exports: Issues for Chile* (SSRN Scholarly Paper ID 3379515). Social Science Research Network. https://papers.ssrn.com/abstract=3379515

Kshetri, N. (2013). Cybercrime and Cybersecurity in Latin American and Caribbean Economies. In N. Kshetri (Ed.), *Cybercrime and Cybersecurity in the Global South* (pp. 135–151). Palgrave Macmillan UK. https://doi.org/10.1057/9781137021946_7

Lehuedé, H. J. (2019). *Corporate governance and data protection in Latin America and the Caribbean* (No. 223; Production Development Series). ECLAC.

MacAskill, E. (2017, November 28). WikiLeaks publishes "biggest ever leak of secret CIA documents" | Media | The Guardian. *The Guardian*. https://www.theguardian.com/media/2017/mar/07/wikileaks-publishes-biggest-ever-leak-of-secret-cia-documents-hacking-surveillance

Malaquias, R. F., & Albertin, A. L. (2018). Challenges for development and technological advancement: An analysis of Latin America. *Information Development*, 0266666918756170. https://doi.org/10.1177/0266666918756170

Mayor, A. (2018). *Amazon.com: Gods and Robots: Myths, Machines, and Ancient Dreams of Technology*. Princeton University Press. https://www.amazon.com/-/es/Adrienne-Mayor-ebook/dp/B07D56CBW1/ref=sr_1_1?__mk_es_US=%C3%85M%C3%85%C5%BD%C3%95%C3%91&dchild=1&keywords=Gods+and+Robots%3A+Myths%2C+Machines%2C+and+Ancient+Dreams+of+Technology&qid=1592881010&s=books&sr=1-1

Mendoza Enríquez, O. A. (2018). Marco jurídico de la protección de datos personales en las empresas de servicios establecidas en México: Desafíos y cumplimiento. *Revista IUS*, *12*(41), 267–291.

Michalczewsky, K., & Ramos, A. (2017). *E-commerce in Latin America: The Regulatory Gap—Conexión Intal* (Ideas de Integración Integration Ideas No. 246). BID. https://conexionintal.iadb.org/2017/03/08/comercio-electronico-en-america-latina-la-brecha-normativa-2/?lang=en

OAS. (2018). *Guidelines for Empowering and Protecting Child and Adolescent Rights on the Internet in Central America and the Dominican Republic*. OAS (Organization of American States).

O'Flaherty, K. (2018, May 3). Cyber Warfare: The Threat From Nation States. *Forbes*. https://www.forbes.com/sites/kateoflahertyuk/2018/05/03/cyber-warfare-the-threat-from-nation-states/

Sarkar, S. (2015). Use of technology in human trafficking networks and sexual exploitation: A cross-sectional multi-country study. *Transnational Social Review*, *5*(1), 55–68. https://doi.org/10.1080/21931674.2014.991184

Schuster, S., van den Berg, M., Larrucea, X., Slewe, T., & Ide-Kostic, P. (2017). Mass surveillance and technological policy options: Improving security of private communications. *Computer Standards & Interfaces*, *50*, 76–82. https://doi.org/10.1016/j.csi.2016.09.011

Segal, A. (2019, December 2). When China Rules the Web. *Foreing Affairs*. https://www.foreignaffairs.com/articles/china/2018-08-13/when-china-rules-web

Sevillano, R., & Rincón, E. G. (2020, April 7). *Los datos de entierros en Madrid apuntan que las muertes por coronavirus pueden ser 3.000 más que las de la estadística oficial*. EL PAÍS. https://elpais.com/sociedad/2020-04-07/los-datos-de-entierros-en-madrid-destapan-hasta-3000-muertes-mas-que-la-estadistica-oficial-de-coronavirus.html

Shillito, M. R. (2019). Untangling the 'Dark Web': An emerging technological challenge for the criminal law. *Information & Communications Technology Law*, *28*(2), 186–207. https://doi.org/10.1080/13600834.2019.1623449

Silva, J., Solano, D., Fernandez, C., Romero, L., & Vargas Villa, J. (2019). *bo Preserving, Protection of Personal Data, and Big Data: A Review of the Colombia Case*. https://repositorio.cuc.edu.co/handle/11323/4836

The Economist. (2018, March). The battle for digital supremacy—America v China. *The Economist*. https://www.economist.com/news/leaders/21738883-americas-technological-hegemony-under-threat-china-battle-digital-supremacy

The Economist. (2020, June 20). Latin America opens up before it's ready. *The Economist.* https://www.economist.com/the-americas/2020/06/20/latin-america-opens-up-before-its-ready

Todt, K. E. (2019). Data Privacy and Protection: What Businesses Should Do. *The Cyber Defense Review, 4*(2), 39–46. JSTOR. https://doi.org/10.2307/26843891

Vila Moreno, M. (2020, March 13). China Adapts to a Changing Latin America. *The Diplomat.* https://thediplomat.com/2020/03/china-adapts-to-a-changing-latin-america/

World Health Organization. (2020). *Coronavirus disease 2019 (COVID-19)* (No. 62; Situation Report). World Health Organization.

Yuan, L. (2020, January 22). China Silences Critics Over Deadly Virus Outbreak. *The New York Times.* https://www.nytimes.com/2020/01/22/health/virus-corona.html

Index

For Product Safety Concerns and Information please contact our EU
representative GPSR@taylorandfrancis.com
Taylor & Francis Verlag GmbH, Kaufingerstraße 24, 80331 München, Germany

www.ingramcontent.com/pod-product-compliance
Lightning Source LLC
Chambersburg PA
CBHW060300220326
41598CB00027B/4173